FLOTATION

MULTISTAGE AND GENERALIZED MODELS OF THE PROCESS

HARVESTERS OF KSENOFONTOV TYPE AND FOR SPECIAL PURPOSE

BORIS S. KSENOFONTOV

Translator:
Darya Dementyeva, PhD in Bioorganic Chemistry

ACADEMUS
Publishing

Academus Publishing
2021

ACADEMUS
Publishing

Academus Publishing, Inc.

1999 S, Bascom Avenue, Suite 700 Campbell CA 95008
Website: www.academuspublishing.com
E-mail: info@academuspub.com

The right of Boris S. Ksenofontov, Doctor of Technical Sciences,
Professor of "Ecology and Industrial Safety" Department, Bauman Moscow State Technical
University, member of Russian Ecology Academy.
For e-mail correspondence: kbsflot@mail.ru

Translator: Darya Dementyeva, PhD in Bioorganic Chemistry.
For e-mail correspondence: dasha160370@icloud.com

ISBN 10: 1 4946 0022 6
ISBN 13: 978 1 4946 0022 8
DOI 10.31519/0022-8

A multistage and generalised flotation model, suggested more than 30 years ago by the author, is considered in a wide aspect for the first time in world literature for reader's attention in monography. The possibilities of its usage are shown in different directions of water flotation purification, sediment thickening and enrichment of minerals. We have shed a light widely on matters concerning new flotation equipment as flotation harvesters of KBS type and for special purposes, which are developed on the basis of flotation process multistage and generalized models. Perspectives and intensification ways of water purification flotation processes are pointed out.

It is suggested for a wide range of readers, including researches, Higher education teachers, PhD students, Masters and Bachelors, Graduate students.

INTRODUCTION

A water purification has been continuing to stay top-priority direction during last decades. Due to this, the development of new ways and devices as well as the improvement of existing ones will represent an undoubtful scientific and practical interest for the wide range of researchers. The special place in new approach development and improvement of existing ones will be given to these processes modelling [1–13].

Flotation technique use in water purification practice can be widen significantly while development of this process scientific basis and working out on this fundament of a flotation technique calculation methodology [1–55]. Thereupon, during more than last three decades, we have been developing a new approach to settle this direction tasks [11–12]. Whereat, not only theoretical basics of this process have been working out but also, apparatus decoration on this new scientific basis has been developing, namely, that is with application of flotation multistage model which we offered in 1987 [13]. The improvement of usage of multistage model further has been pursued almost for all utilized in practice flotation ways of water purification as well as in the processes of wastewater sediment thickening, including ones of excessive active silt and, partially, in mineral enrichment technology. The use of multistage and generalized models has changed the scientific approach in flotation technology from fragmental to continuous one, beginning from conditioning of being cleaned wastewaters and being thickened sediments and ending with finish stages of wastewater polishing and sediment dehydration [50].

It's worth to mention especially the development on this scientific base of a flotation equipment — flotation harvesters, including ones of KBS-type and for special purpose wherein not only water purification but preliminary sediment thickening is pursued [33–35, 52–62]. It should be mentioned in a whole that the use of scientific base during its realization will allow essentially to raise water cleaning efficiency and settle the problem of preliminary thickening of wastewater sediments.

The author hopes that with correct usage of multistage model, flotation technology of wastewater purification, sediment thickening and mineral enrichment will become noticeably more effective.

Meanwhile, another the most important task is the use of a generalized multistage flotation model for to calculate the processes of extraction from water of not only hydrophobic but also hydrophilic pollutions. The

3

application of generalized multistage flotation model has given us a possibility to develop both harvesters, of general KBS-type and for special purpose. These new technique kinds for water purification give opportunity to reach higher technological results in comparison with the use of more known facilities of analogous type.

1 MULTISTAGE MODEL OF KSENOFONTOV FLOTATION AND ITS APPLICATION

Rather many models of flotation process are known [11–12, 36–47], but especially the model of Professor Beloglazov should be mentioned. According to this model, flotation process is considered similarly to simple chemical reaction of the first order (Fig. 1.1)

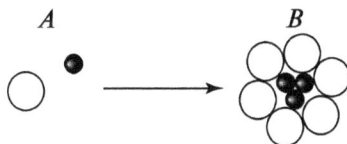

Fig. 1.1. Scheme of Beloglazov flotation scheme

Equation of Beloglazov flotation process has the following form:

$$C = C_0 e^{-k\tau},$$ (1.1)

where C and C_0 — contamination concentrations at current and initial moments, correspondingly;

k — constant, characterisinng floatation proccess speed;

τ — time.

$$k = \frac{1.5qE}{k_0 \overline{D}},$$ (1.2)

where q — barbotage speed;

E — efficiency of particles capture by floating gas bubble while a flotation;

k_0 — factor of bubbles poly-dispersion;

$\overline{D}$ — bubble average diameter in a flotation cell.

Essential drawbacks of this approach are:
- floto-complex is not considered as a research subject;
- there is no dependency of foam product formation versus a time;
- factors, influencing process kinetics, are not fully indicated.

Elimination of these drawbacks and the fullest description of flotation process had been offered by B.S. Ksenofontov in the middle of 80s of C20th [11–13, 36–47, 49–52]. According to the approach, a flotation process is considered similarly to complex chemical reaction of the first order.

It's worth to mention that the development of this model by the author began from the discussion in 70s–80s of C20th between the author and Professor Klassen who was affirming that Beloglasov model described well the experimental data. Nevertheless, the author's firm convincement was and continues to be inflexible about that flotocomplex particle-bubble should be at the basis of flotation process model. It had been put as a ground for proposed model that was first published by Ksenofontov B.S. in 1987 [13]. In different author discussions with adversaries, model various individual cases would be considered which used to be proposed especially by young researchers who would claim the approach originality, but then it was clarified that all these were sub-cases of the author's model because the major feature — flotocomplex particle-bubble existence — was at their basis. Possible cases of the author's model with the consideration of not only model major feature, flotocomplex particle-bubble, but of others, for instance, reversibility of flotation process separate stages and others will be considered further during the material presentation. Together, according to the author, tight analogy between the proposed model and complex chemical reaction is seen where major object is an intermediate product (complex). The author research during more than 30 years has confirmed this convincingly.

Flotation process simplest case due to Ksenofontov model is shown on Fig. 1.2, and more general case — on Fig. 1.3.

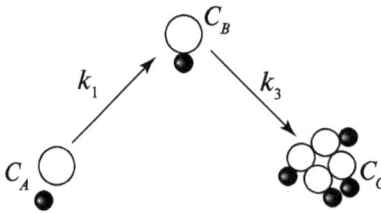

Fig. 1.2. Flotation process
simplest case according to
Ksenofontov model

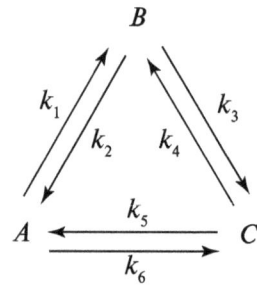

Fig. 1.3. General case of flotation
process according
to Ksenofontov model

Equation system for the simplest case has the following form:

$$
\begin{cases}
\dfrac{dC_A}{dt} = -k_1 C_A \\[2mm]
\dfrac{dC_B}{dt} = k_1 C_A - k_3 C_B \, . \\[2mm]
\dfrac{dC_c}{dt} = k_3 C_B
\end{cases}
\tag{1.3}
$$

Equation system for general case has the following form:

$$
\begin{cases}
\dfrac{dC_A}{dt} = -k_1 C_A + k_2 C_B + k_5 C_C - k_6 C_A; \\[2mm]
\dfrac{dC_B}{dt} = k_1 C_A - k_2 C_B - k_3 C_B + k_4 C_C; \\[2mm]
\dfrac{dC_c}{dt} = -k_5 C_C + k_6 C_A + k_3 C_B - k_4 C_C,
\end{cases}
\tag{1.4}
$$

$$
k_1 = \frac{1.5qE}{k_0 \overline{D}},
\tag{1.5}
$$

where q — barbotage speed;

E — efficiency of particles capture by floating gas bubble while a flotation;

k_0 — factor of bubbles poly-dispersion;

$\overline{D}$ — bubble average diameter in a flotation cell.

The definition and calculation of constants $k_2 - k_6$ were first given by B.S. Ksenofontov [13, 14].

$$
k_2 = A C_f G_a M^2 C_u f^{(-1)},
\tag{1.6}
$$

where A — dimensionless coefficient;

C_f — concentration of flotocomplexes particle-bubble;

C_a — speed gradient in aeration zone, defined by ratio of speed differences to difference of distances between points being-considered;

M — ratio of particle diameter to bubble diameter;

$C_u f$ — bubble concentration in underfoam layer.

$$k_3 = \frac{\upsilon_{lift}}{h}, \tag{1.7}$$

where υ_{lift} — flotocomplex raise speed;

h — distance from aeration zone to foam layer

$$k_4 = FG_uC_bd_{av}^3, \tag{1.8}$$

where F — ptoportionality coefficient;

G_u — speed gradient in underfoam layer;

C_b — bubble concentration in foam;

d_{av} — bubble avearage diameter in a foam

$$k_5 = \frac{\upsilon_{sed}}{h}, \tag{1.9}$$

where υ_{sed} — sedimentation speed of particles in solid phase, for particles falling from foam layer;

h — distance from aeration zone to foam layer

$$k_6 = \psi \frac{\partial}{\partial x} \left\{ \frac{1}{2\sqrt{\pi\psi t}} \left[\exp\left(-\frac{(x-h)^2}{4\psi t} \right) - \exp\left(-\frac{(x+h)^2}{4\psi t} \right) \right] \right\}, \tag{1.10}$$

where t — time;

χ — current distance from foam layer border;

ψ — diffusion coefficient of particles of solid phase into a liquid;

h — distance from aeration zone to foam layer.

The solution of equation system (1.4) has the form, presented on Fig. 1.4.

Flotation process intensification according to a multistage model can be reached in a row of ways, including flotocomplex coalescence way with the formation of the bubble of larger size than the original one (Fig. 1.5).

Equation system, describing this case, has the following form (1.11):

$$\frac{dC_A}{dt} = -k_1C_A + k_2C_B + k_5C_C - k_6C_A;$$

$$\frac{dC_B}{dt} = k_1C_A - k_2C_B - k_3C_B + k_4C_C - k_7C_B + k_8C_D;$$

$$\frac{dC_c}{dt} = -k_5C_C + k_6C_A + k_3C_B - k_4C_C + k_9C_D - k_{10}C_C; \tag{1.11}$$

$$\frac{dC_D}{dt} = k_7C_B - k_8C_D - k_9C_D + k_{10}C_C.$$

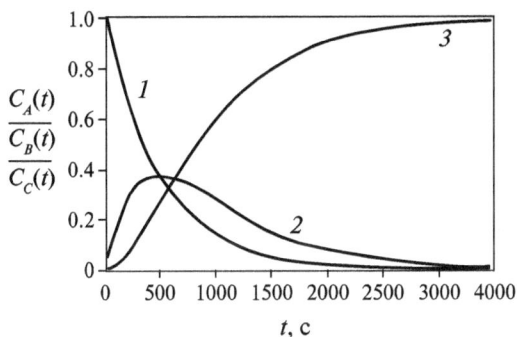

Fig. 1.4. Graphical solution of equation system of flotation process according to Ksenofontov model.
Pollution concentration dependence:
1 — in the liquid, being defecated; 2 — in the form of flotocomplexes; 3 — in foam product

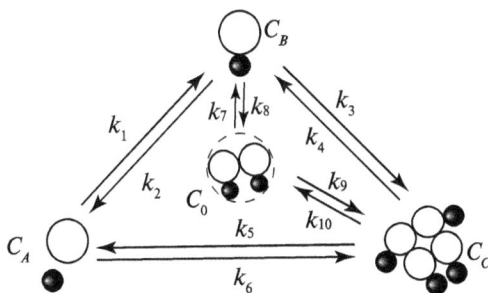

Fig. 1.5. Flotation scheme taking into account coalescence phenomenon of pollution particles

This system solution in graphical form is demonstrated on Fig. 1.6 and 1.7.

Other various variants of the intensification of the flotation process with the use of multistage model were considered in our particular works and described the flotation process being pursued in concrete conditions [36–47].

For example, the description of the processes of ionic flotation on the basis of multistage model [32, 42] can be presented in the form of sequence of the following system states (Fig. 1.8):

- state A — the ions of coligand and collector and gas bubbles exist anonymously;
- state B — sublat formation as a result of interaction between collector and coligand;

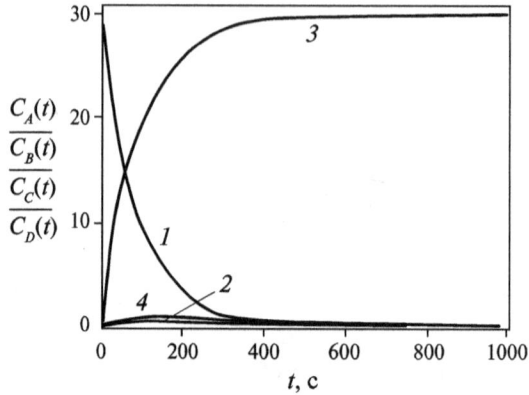

Fig. 1.6. Dependence of change of hydrophobic pollution (mineral oil products) concentrations versus flotation time taking into account coalescence. Pollution concentration dependence:
1 — in the liquid, being defecated; *2* — in the form of flotocomplexes; *3* — in foam product;
4 — in micro-flotocomplex state

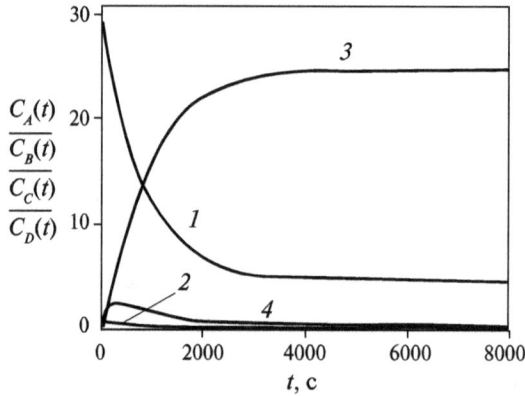

Fig. 1.7. Dependence of change of suspended substances concentrations versus flotation time taking into account coalescence. Pollution concentration dependence:
1 — in the liquid, being defecated; *2* — in the form of flotocomplexes; *3* — in foam product;
4 — in microflotocomplex state

- state C — the formation of flotocomplex collector-(gas (air) bubble);
- state D — the formation of flotocomplex (coligand ion)-collector-(gas bubble);

- state E — the formation of foam layer, containing the ions of coligand and of collector and the gas bubbles;
- state F — the formation of foam, containing the ions of kolligends and of collector without the gas bubbles (sublat concentrate).

The mathematical description of flotation process, presented on Fig. 1.8, can be demonstrated by following equation system:

$$\begin{cases} \dfrac{dC_A}{dt} = -k_1C_A + k_2C_B - k_3C_A + k_4C_C - k_{15}C_A + k_{16}C_F; \\[2mm] \dfrac{dC_B}{dt} = k_1C_A - k_2C_B - k_5C_B + k_6C_D - k_{13}C_B + k_{14}C_F; \\[2mm] \dfrac{dC_C}{dt} = k_3C_A - k_4C_C - k_7C_C + k_8C_D; \\[2mm] \dfrac{dC_D}{dt} = k_5C_B - k_6C_D + k_7C_C - k_8C_D - k_9C_D + k_{10}C_E; \\[2mm] \dfrac{dC_E}{dt} = k_9C_D - k_{10}C_E - k_{11}C_E + k_{12}C_F; \\[2mm] \dfrac{dC_F}{dt} = k_{11}C_E - k_{12}C_F + k_{13}C_B - k_{14}C_F + k_{15}C_A - k_{16}C_F. \end{cases} \qquad (1.12)$$

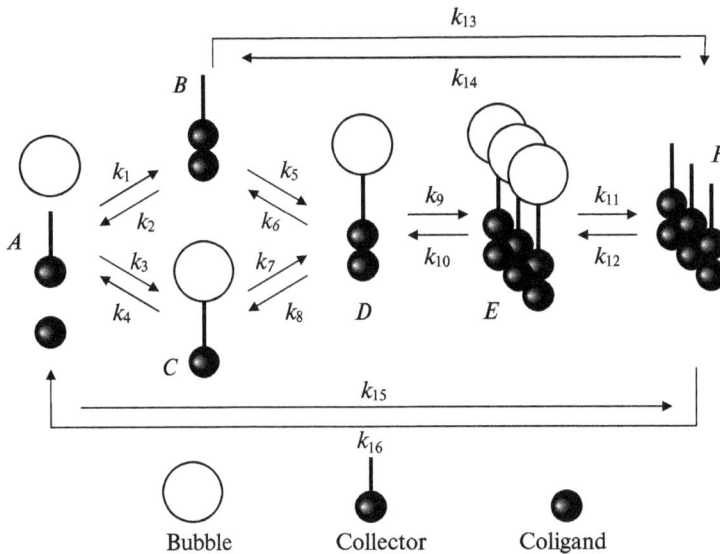

Fig. 1.8. Scheme of ionic flotation multistage model

The proposed system should satisfy, at least, two conditions, namely, at the starting moment, coligand concentration on the first stage is equal to the original concentration in solution, and at any time moment, the sum of coligand concentrations by all stages is equal to its original concentration.

The use of flotation multistage model allows to approach scientifically validly the development of new kinds of flotation technique.

One of the major directions of flotation technique development — is the creation of larger and more rentable machines. Recently, the transition from the principles of simplification of constructions and mechanisms to the principles, which allow to separate and to direct liquid streams, to provide for air external inflow towards large devices and, thus, to use monochambers of big volume, has taken place.

2 FLOTATION PROCESS KINETICS ON MULTISTAGE MODEL BASIS

2.1 FLOTATION PROCESS KINETICS ON THE EXAMPLE OF IONIC FLOTATION

The consideration of ionic flotation process in the frames of equation system (1.12) demands as a rule the application of calculation methods.

For practical cases, as our calculations have shown, the simplified calculations can be used well, including with the use of the scheme, presented on Fig. 2.1.

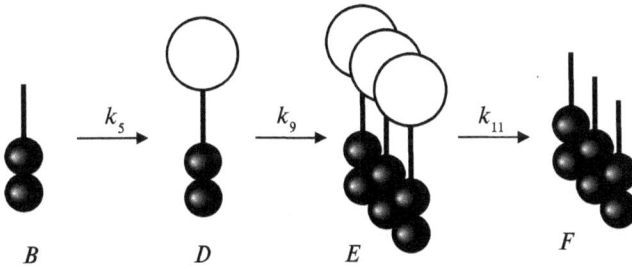

$$B \xrightarrow{\ k_5\ } D \xrightarrow{\ k_9\ } E \xrightarrow{\ k_{11}\ } F$$

Fig. 2.1. Simplified scheme of ionic flotation process.

Considering that collector has completely reacted with coligand, equation system will take the form (2.1):

$$\begin{cases} \dfrac{dC_B}{dt} = -k_5 C_B; \\[2mm] \dfrac{dC_D}{dt} = k_5 C_B - k_9 C_D; \\[2mm] \dfrac{dC_E}{dt} = k_9 C_D - k_{11} C_E; \\[2mm] \dfrac{dC_F}{dt} = k_{11} C_E. \end{cases} \tag{2.1}$$

The efficiency of individual metals extraction from wastewaters, obtained by calculation and defined experimentally, is presented in Table 2.1.

Table 2.1. Extraction indicators of individual metals from wastewater with the use of ionic flotation

№ n/n	Metal	Metal concentration in wastewater, mg/l	Flotation time, min	Calculated purification efficiency, %	Purification efficiency determined experimentally, %
1	Chrome (general)	2.2	15.5	97.7	91.4
2	Lead	4.4	15.5	96.9	89.6
3	Nickel	3.5	15.5	98.4	92.7
4	Wolfram	2.9	15.5	97.8	93.3
5	Cobalt	4.1	15.5	96.5	89.6

Meanwhile, the biomass of the bacteria of *Pseudomonas* kind with 0.7 g/l dose was used as a collector, together, bacteria biomass had been led to disintegration to separate fragments of bacteria cells before the introduction into the water.

Comparison of calculated and experimental values of efficiency of individual metal ion extraction from wastewaters shows a little difference, approximately not exceeding 7%, that allows to use calculated data for efficiency evaluation of wastewater purification, including metals, with the application of ionic flotation.

2.2 KINETICS OF WASTEWATER PURIFICATION PROCESS BY FLOTATION DEFECATING

During works conducted by us last years, it has been established that combined flotation devices — flotation harvesters appear in the row of cases to be more effective than the use of apparatuses wherein only one water purification process is used as flotation, defecating, filtration or `so forth. Theoretical analysis and experimental data show that, in the new case, material and energetic expenditure decline, and, besides, the efficiency of purification here exceeds the additive value summarized from the effects of flotational, sedimentation and filtration technologies of purification. For to embody such approach we are carrying out new types of combined technique on the basis of flotation machines and apparatuses. These types of equipment have got the title of flotoharvester, introduced in a row of enterprises. The modelling of purification processes on these apparatuses is feasible in more effective application of such technique in wastewater purification practice. Let consider on the example of the model for co-joint process including the flotation and parallelly proceeding defecating in flotation decanter, presenting the simplest flotoharverster type.

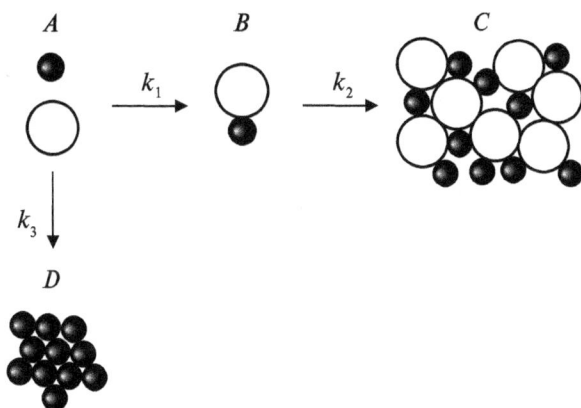

Fig. 2.2. Scheme of flotodefecating process without accounting for reversibility

Flotodefecating process models without reagent usage or with reagents have been worked out on the basis of B.S. Ksenofontov's flotation multistage model. The scheme of the process, proceeding in flotodecanter of column type, in the simplest form, without reagent application and account for the process reversibility, is shown on Fig. 2.2.

Given process is described by the following differential equation system:

$$\frac{dC_A}{dt} = -k_1 C_A - k_3 C_A;$$

$$\frac{dC_B}{dt} = k_1 C_A - k_2 C_B;$$

$$\frac{dC_c}{dt} = k_2 C_B$$

$$\frac{dC_D}{dt} = k_3 C_A,$$

where A — particle initial state;

B — particle state of sticking and fixing on bubbles;

C — particle state in foam layer;

D — state of particles, fallen out into sediment;

CA, CB, CC and CD — particle concentration in states A, B, C and D, correspondingly;

k_1, k_2 и k_3 — constants of particle transition from one state to another.

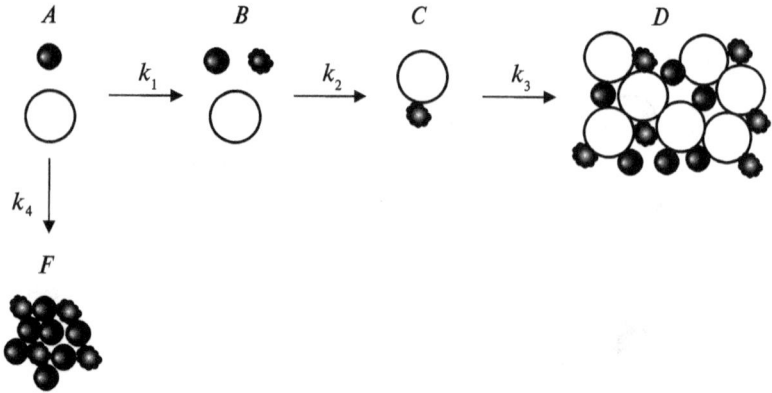

Fig. 2.3. Scheme of flotation defecating process with the use
of reagents without accounting for processes reversibility

To settle shown above differential equation system with initial conditions $t = 0$ $CA = C_0$, $CB = 0$, $CC = 0$, $CD = 0$ software complex Maple 15 was used.

The solution of differential equation system, describing flotodefecating process without taking into account the process reversibility, looks in the following way in analytical form:

$$C_A(t) = C_0 e^{-(k_1+k_3)t};$$

$$C_B(t) = \left(1 - e^{-(k_1-k_2+k_3)t}\right) \frac{k_1 C_0 e^{-k_2 t}}{k_1 - k_2 + k_3};$$

$$C_C(t) = -\left(\frac{1}{k_2} - \frac{e^{-(k_1-k_2+k_3)t}}{k_1+k_3}\right) \frac{k_1 k_2 C_0 e^{-k_2 t}}{(k_1 - k_2 + k_3)} + \frac{k_1 C_0}{k_1+k_3};$$

$$C_D(t) = \left(1 - e^{-(k_1+k_3)t}\right) \frac{k_3 C_0}{k_1+k_3}.$$

Very often, for wastewater purification, reagents are used which promote flotation process intensification. The scheme of reagent-kind flotation defecating model without taking into account reversibility of processes is shown of Fig. 2.3.

In this case, purification process is described by the following differential equation system:

$$\frac{dC_A}{dt} = -k_1 C_A - k_4 C_A;$$

$$\frac{dC_B}{dt} = k_1 C_A - k_2 C_B;$$

$$\frac{dC_C}{dt} = k_2 C_B - k_3 C_C;$$

$$\frac{dC_D}{dt} = k_3 C_C;$$

$$\frac{dC_F}{dt} = k_4 C_A,$$

where A — particle initial state;

B — state of particles interaction with reagents;

C — particle state of sticking and fixing on bubbles;

D — particle state in foam layer;

F — state of particles fallen out into sediment;

CA, CB, CC, CD and CF — particle concentration in states A, B, C, D and F, correspondingly;

k_1, k_2, k_3 and k_4 — constants of particle transition from one state to another.

To settle differential equation system, describing flotodefecating process with the reagent use, the software complex Maple 15 was used.

Solving differential equation system, describing flotodefecating process with reagents usage, at initial conditions $t = 0$ $CA = C_0$, $CB = 0$, $CC = 0$, $CD = 0$, $CF = 0$ we get the following in analytical form:

$$C_A(t) = C_0 e^{-(k_1+k_4)t};$$

$$C_B(t) = \left(1 - e^{-(k_1-k_2+k_4)t}\right) \frac{k_1 C_0 e^{-k_2 t}}{k_1 - k_2 + k_4};$$

$$C_C(t) = \frac{k_1 k_2 C_0}{k_1 - k_2 + k_4} \left(\frac{e^{-(k_1+k_4)t}}{k_1 - k_2 + k_4} + \frac{(k_1 + k_4)e^{-k_2 t}}{(k_1 - k_2 + k_4)(k_3 - k_2)} \right);$$

$$C_D(t) = -\frac{k_1 k_2 k_3 C_0 e^{-(k_1+k_4)t}}{(k_1 - k_2 + k_4)(k_1 - k_3 + k_4)(k_1 + k_4)} + \frac{k_1 k_2 k_3 C_0 e^{-k_2 t}}{(k_1 - k_2 + k_4)(k_2 - k_3)} +$$

$$+\frac{k_1 k_2^2 C_0 e^{-k_3 t}}{(k_1 - k_3 + k_4)(k_2 - k_3)};$$

$$C_F(t) = \left(1 - e^{-(k_1 + k_4)t}\right)\frac{k_4 C_0}{k_1 + k_4}.$$

Obtained solutions were used in the management of drainages purification processes in combined flotation devices.

More complex cases of above described processes deserve special attention, particularly, with accounting for process individual stage reversibility. The scheme of such combined process, conjoing flotation and defecating, is shown on Fig. 2.4.

Differential equation system in this case has the form (2.2):

$$
\begin{cases}
\dfrac{d}{dt}C_A = -k_1 C_A + k_2 C_B - k_6 C_A; \\[2mm]
\dfrac{d}{dt}C_B = k_1 C_A - k_2 C_B - k_3 C_B + k_4 C_C; \\[2mm]
\dfrac{d}{dt}C_C = k_3 C_B - k_4 C_C - k_5 C_C; \\[2mm]
\dfrac{d}{dt}C_D = k_6 C_A + k_5 C_C;
\end{cases}
\qquad
\begin{aligned}
& t = 0: \\
& C_A(0) = C_{A_0}; \\
& C_B(0) = 0; \\
& C_C(0) = 0.
\end{aligned}
\qquad (2.2)
$$

Finding of pollution values at any time moment can be accomplished from equation system solutions (2.2) at above specified initial conditions.

This equation system solution has the form (2.3):

$$
\begin{aligned}
C_A(t) &= f(k_1 \dots k_6); \\
C_B(t) &= f(k_1 \dots k_6); \\
C_C(t) &= f(k_1 \dots k_6); \\
C_D(t) &= f(k_1 \dots k_6).
\end{aligned}
\qquad (2.3)
$$

The obtained solution analysis points out solution complexity and some difficulty in its interpretation. Nevertheless, it represents undoubtful interest for further practical tasks with the aim to optimize linked processes of flotation and defecating.

Let consider the water purification process with the use of flotation and accompanying it defecating without reversibility of individual stages of this process representing big practical interest (Fig. 2.5)

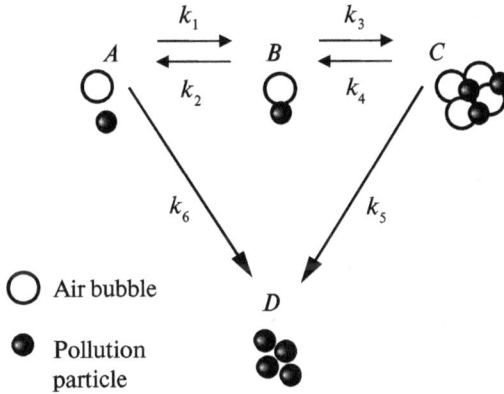

Fig. 2.4. Scheme of combined process of flotation and defecating

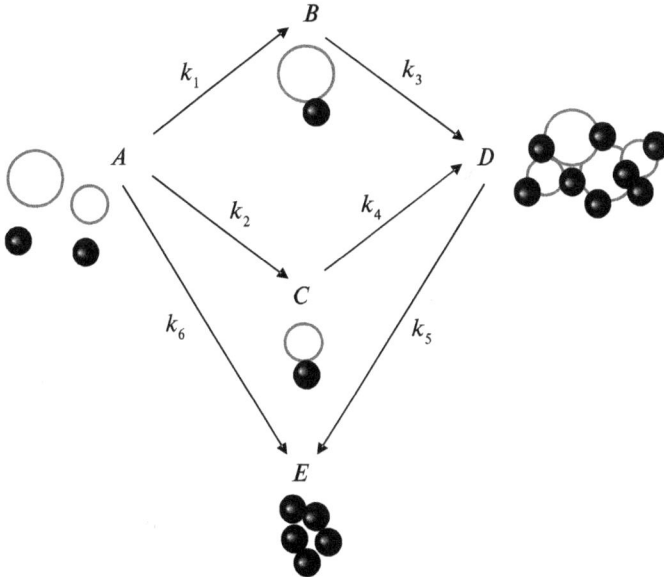

Fig. 2.5. Scheme of combined processes of flotation and defecating

In the given scheme, A — pollution particles are in initial state. During flotation cleaning, the particles of hydrophobic pollutions can stick to bigger bubbles, forming flotocomplexes (state B), and particles of hydrophobic-hydrophilic pollutions form flotocomplexes with bubbles of smaller sizes (state C). The flotocomplexes emerge on the surface forming the foam layer (state D). At once with flotocomplexes emergence, sedimentation of particles can proceed (state E) from initial state A as well as from foam layer D. The transitions between states are characterized by constants k_1, k_2, k_3, k_4, k_5, k_6.

Definition of constants certain values and their maintenance on certain numerical level in the exploitation process can optimize wastewater purification regime. Calculation of constants was held by formulas (1.5–1.10).

Let's accept average barbotage speed as $q = 1.6 \cdot 10^{-5}$ m^3/(m^2·s).

The efficiency of particle capture by surface-emerging (floating) bubbles of gas during flotation is accepted as $E = 0.05$; bubble polydispersity factor is accepted as $k_0 = 1$; average diameter for large bubble — $D_1 = 500$ microns, and for small bubble — $D_2 = 60$ microns, so we get following values:

$$k_1 = \frac{1.5 \cdot 1.6 \cdot 10^{-5} \cdot 0.05}{1 \cdot 500 \cdot 10^{-6}} = 0.0024 \, \frac{1}{s};$$

$$k_2 = \frac{1.5 \cdot 1.6 \cdot 10^{-5} \cdot 0.05}{1 \cdot 60 \cdot 10^{-6}} = 0.02 \, \frac{1}{s}.$$

It is known from experimental data: surface-emerging speed for bubbles of 500 microns diameter: $v_{em1} = 0.07$ m/s; surface-emerging speed for bubbles of 60 micron diameter: $v_{em2} = 0.0019$ m/s.

In the case of usage of flotocell with working depth of 1 m (h = 1m), the values of k_3, k_4 constants will be according to (1.7) equal to:

$$k_3 = \frac{0.07}{1} = 0.07 \, \frac{1}{s};$$

$$k_4 = \frac{0.0019}{1} = 0.0019 \, \frac{1}{s}.$$

k_5, k_6 constants, characterizing sedimentation process, are being determined according to particle sedimentation speed $v_{sed} = 0.2 \cdot 10^{-3}$ m/s;

$$k_5 = k_6 = \frac{0.2 \cdot 10^{-3}}{1} = 0.0002 \, \frac{1}{s}.$$

Graphs of concentration change dependencies from time (Fig. 2.6) are made with the use of calculated constants on the basis of particular data according to found solution.

Varying different parameters, we can establish what extent to, their change influences purification process.

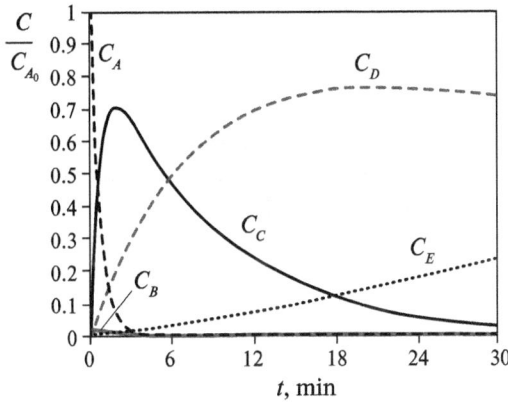

Fig. 2.6. Pollution concentration dependence from process time (for bubbles of diameter of 500 and 60 microns)

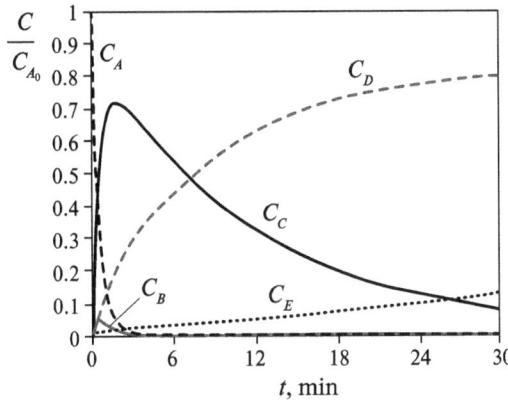

Fig. 2.7. Pollution concentration dependence from process time (for bubbles with 300- and 50-micron diameters)

Fig 2.7 shows solutions in graphical form for the following data:
D_1 = 300 microns; D_2 = 50 microns;
v_{sed1} = $0.1 \cdot 10^{-3}$ m/s — speed of particle sedimentation from foam;
v_{sed2} = $0.5 \cdot 10^{-3}$ m/s — speed of particle sedimentation from initial state;
k_1 = 0.004 1/s; k_2 = 0.024 1/s; k_3 = 0.035 1/s; k_4 = 0.0013 1/s;
k_5 = 0.0001 1/s; k_6 = 0.0005 1/s.

Presented data show how values of flotation process constants influence kinetics of the process. Considered model demonstrates characteristic specialties of multistage process wherein sedimentation proceeds

21

besides the flotation. The kind of curves (Fig. 2.6–2.7) demonstrates how the change of characteristic parameters determines the process flow on individual stages and in a whole: for instance, on Fig. 2.6, C_D curve, characterizing particles concentration in foam layer, starts to decline after 24 minutes. Together, it can be concluded that particle falling out from foam layer, beginning from this time moment, proceeds more intensively than surface-emerging, but such decline has not to be observed at other values of parameters (Fig. 2.7).

Thus, obtained dependencies give possibility to define purification process lasting. The process is usually ceased when concentration change versus time becomes constant or the change doesn't exceed 5% because, in this case, the change is comparable with measurement bial. Nevertheless, it is necessary to take into account the process characteristic specialties in certain conditions and normative demands, claimed on water quality, in each individual case while purification time determination [32].

3 MODELLING OF WASTEWATER ELECTRO–FLOTATION PURIFICATION

3.1 ELECTRO-FLOTATION PROCESS THEORETICAL BASICS

Electro-flotation technique has been used for wastewater purification during decades [22, 23]. Nevertheless, the problems of this method wide usage still exist. The research, that we have pursued approximately during 30 years, on clarification of possibilities of wastewater purification electro-flotation way application has shown its usage efficiency only in individual cases when other technologies do not have any effect. The electro-flotation technique usage is also pointed out in the world experience, also including leading companies in wastewater purification sphere. Whereat, it should be noted that, in the case of usage of electro-flotation technique of low working efficiency (about 1...5 m³/h), the obtaining of this method opponent advantages in comparison of other methods is possible. Pursued by us the unique industrial experiment on the trial of electronic flotation slimmer of 100 m³/h working efficiency has shown that, in small production conditions, such apparatus exploitation is complicated because of electro-energy high expenditure. Electrical capacity of such apparatus is about 150...300 kw depending on wastewater qualitive content, including its specific conductivity. These trials have confirmed that electro-flotation apparatus of such productivity shouldn't be applied in the majority of cases in wastewater purification processes. It is profitable to apply electric flotation skimmers in some cases while usage of low productivity devices and especially of ones of lower than 1 m³/h productivity. Our experience of introduction on different enterprises of Russia and Finland has confirmed it.

Further development of electro-flotation technique is connected much with this approach theoretical basics as well as with the usage of new materials for electrodes, the improvement of electric flotation skimmers designs, including ones of the combined type. Particular developments of this type apparatuses have been led by us till industrial introduction. Nevertheless, this method theoretical grounds on the basis of general views on flotation process are still absent that hinders its improvement.

To manage and optimize this approach it is necessary to know electro-flotation process theoretical grounds and to model it. Some important theoretical evidences, proposed in a series of works, do not give comprehensive evaluation of electro-flotation approach on the grounds of basic knowledge of flotation process. Started by us 25 years ago systematic investigations on flotation theory established new basis for the new approach which is founded on the usage of multistage model of flotation processes, also including electro-flotation one [13, 32]. According to this model, any flotation process is characterized by several constants which the most important from are k_1, characterizing the formation of flotocomplex (pollution particle) — (gas (air) bubble), and constant k_3, characterizing the raise of these flotocomplexes up into a foam layer.

For electro-flotation process, flotocomplex formation probability, characterized by k_1 constant, can be determined with the use of relation (3.1):

$$k_1 = \frac{1.5qE}{k_0\overline{D}},\qquad(3.1)$$

where q — barbotage speed, m³/(m²·s);

E — efficiency of particle capture by surface-emerging oxygen or hydrogen bubbles during electro-flotation;

$\overline{D}$ — average diameter of oxygen and hydrogen bubbles in flotation cell, m;

k_0 — polydispersity factor of bubbles (dimensionless).

Flotocomplex surface-emerging, characterized by k_3 constant, is defined by formula (3.2):

$$k_3 = \frac{\upsilon_{em}}{h},\qquad(3.2)$$

where υ_{em} — flotocomplex emerging speed, m/s;

h — distance from aeration zone till foam layer (flotochamber depth), m.

During electro-flotation, aeration intensity q and diameters D of formed bubbles depend on current density j.

Let's define the dependency of aeration intensity from current density. Aeration intensity is defined by following formula (3.3):

$$q = \frac{Q}{S},\qquad(3.3)$$

where Q — expenditure of gases (oxygen and hydrogen), m³/s;

S — square of aeration chamber cross-section.

Let's present gas expenditure Q in the form (3.4):

$$Q = \frac{V}{t},$$ (3.4)

where V — gas volume, m^3;
t — time, s.

So, taking into account (3.16), expression (3.15) will look in the following way (3.5):

$$q = \frac{V}{S \cdot t}.$$ (3.5)

According to the Faraday law, mass of the substance, being extracted on electrode during electrolysis, is directly proportional to force of the current I, going through electrolyte with time t (3.6):

$$m = k_e q_e = k_e I t,$$ (3.6)

where k_e — substance electrochemical equivalent, kg/C (Coloumb);
$\qquad I$ — current force, A;
$\qquad m$ — mass, kg.

Current strength I is connected with current density j by relation (3.7):

$$I = jS.$$ (3.7)

Let's express mass in terms of density and volume, so we'll get (3.8):

$$\rho V = k_e jSt$$ (3.8)

Taking into account (3.8), aeration intensity q can be expressed in the terms of the following formula:

$$q = \frac{k_e j}{\rho}$$ (3.9)

Then, formula for k_1 definition, taking into account (3.9), looks in the following way (3.10):

$$k_1 = \frac{1,5 k_e j E}{k_0 \overline{D} \rho},$$ (3.10)

Gas bubbles are formed during flotation by the way of water electrolysis, whereat hydrogen is extracted on the cathode, and oxygen — on the anode.

Let's consider 3 variants of process flow: elecroflotation on hydrogen, elecroflotation on oxygen, elecroflotation on hydrogen and oxygen.

3.2 ELECTRO-FLOTATION ON HYDROGEN

The scheme of flotation machine, where electro-flotation process proceeds on the hydrogen, is presented on Fig. 3.1.

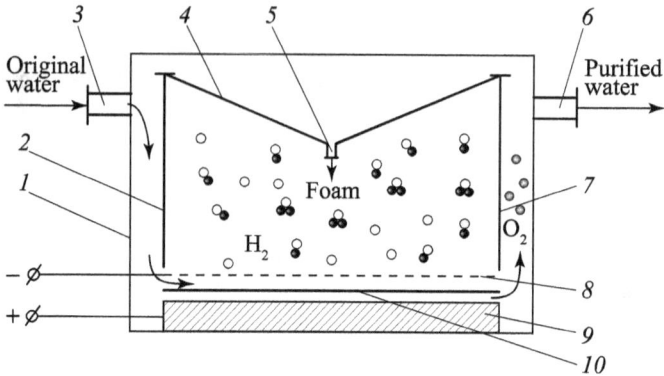

Fig. 3.1. Scheme of the flotation machine:
electro-flotation on hydrogen:
1 — shell; *2, 7* — semi-submersible walls, *3* — water supply sleeve;
4 — foam gutter; *5* — foam bend sleeve; *6* — water bend sleeve;
8 — cathode; *9* — anode; *10* — tarp wall

Original water is supplied along Sleeve *3* and, going through semi-submersible Wall *2*, gets into aeration zone. Hydrogen bubbles are extracted on Cathode *8*; the bubbles, being merged with pollutions, form floto-complexes. Flotocomplexes emerge on a surface forming a foam. The foam gets into foam Gutter *4* and is bent along Sleeve *5*. Tarp Wall *10* hinders oxygen bubble, extracted on Anode *9*, from getting into aeration zone. These bubbles are bent into the zone behind semi-submersible Wall *7*. In this case, the process proceeds according to the scheme presented on Fig. 2.3.

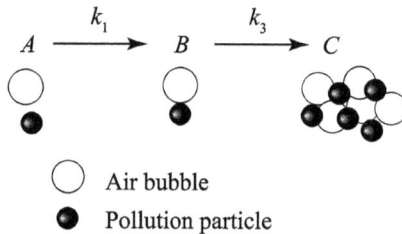

⬤ Pollution particle

Fig. 3.2. Scheme of electro-flotation process on hydrogen

In the given scheme: A — pollution particles in initial state, B — flotocomplexes with hydrogen bubbles, C — state of bubbles in foam layer; constant k_1 characterizes probability of flotocomplex formation during interaction of pollution particle with hydrogen bubble, k_3 characterizes the given flotocomplex surface-emerging.

Differential equation system in this case will look in the following way (3.11):

$$\begin{cases} \dfrac{dC_A}{dt} = -k_1 C_A \\[2mm] \dfrac{dC_B}{dt} = k_1 C_A - k_3 C_B. \\[2mm] \dfrac{dC_C}{dt} = k_3 C_B \end{cases} \qquad (3.11)$$

Initial conditions for given differential equation system at $t = 0$ are the following (3.12):

$$\begin{aligned} C_A(0) &= C_{A_0}; \\ C_B(0) &= 0; \\ C_C(0) &= 0. \end{aligned} \qquad (3.12)$$

Solution of equation system (3.11) will have the following form (3.13):

$$C_A(t) = C_{A_0} e^{-k_1 t};$$
$$C_B(t) = \frac{C_{A_0} k_1}{k_3 - k_1}\left(e^{-k_1 t} - e^{-k_3 t}\right);$$
$$C_C(t) = C_{A_0}\left(1 - \frac{k_3}{k_3 - k_1} e^{-k_1 t} - \frac{k_1}{k_1 - k_3} e^{-k_3 t}\right). \qquad (3.13)$$

According to (3.10), let's calculate constant k_1 for the following values:
- let's accept current density as $j = 10$ mA/sm^2 = 100 A/m^2;
- hydrogen chemical equivalent as $k_e \approx 1{,}045 \cdot 10^{-8}$ kg/C;
- hydrogen density as $\rho = 0{,}09$ kg/m^3;
- let's accept bubble average diameter as $D = 50$ microns according to [56] for $j = 100$ A/m;

- let's accept particle capture efficiency by surface-emerging gas bubble during flotation as $E = 0.05$;
- let's accept bubble polydispersity factor as $k_0 = 1$;

$$k_1 = \frac{1.5 \cdot 1.045 \cdot 10^{-8} \cdot 100 \cdot 0.05}{1 \cdot 50 \cdot 10^{-6} \cdot 0.09} = 0,0174 \frac{1}{s}.$$

We can calculate constant k_3 according to formula (3.2), whereat surface-emerging speed can be approximately determined by formula (3.14):

$$v_{em} = \frac{\bar{D}^2 g (\rho_w - \rho_a)}{18\mu}; \qquad (3.14)$$

$$v_{em} = \frac{(50 \cdot 10^{-6})^2 \cdot 9.81(1000 - 0.09)}{18 \cdot 10^{-3}} = 0.00136 \text{ m/s};$$

- let's accept height as $h = 1$ m;

$$k_3 = \frac{0.00136}{1} = 0.00136 \frac{1}{s}.$$

Dependence charts of concentration versus process time are built with the use of obtained constants on the basis of found solution (Fig. 3.3).

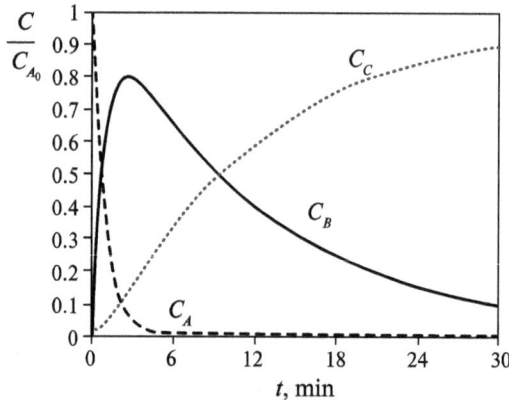

Fig. 3.3. Concentration dependence versus process time during electro-flotation on hydrogen

3.3 ELECTRO-FLOTATION ON OXYGEN

The scheme of flotation machine, where electro-flotation process proceeds on the oxygen, is presented on Fig. 3.4.

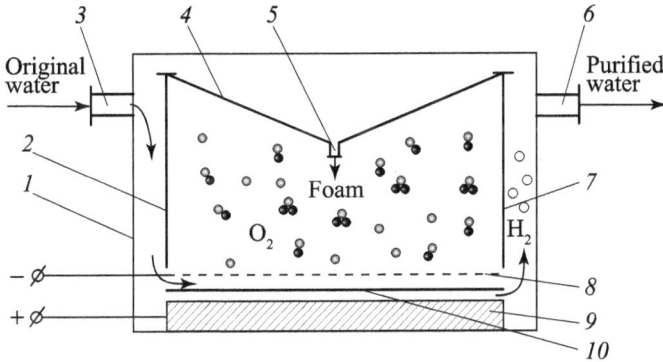

Fig. 3.4. Scheme of the flotation machine:
electro-flotation on oxygen:
1 — shell; *2,7* — semi-submersible walls, *3* — water supply sleeve;
4 — foam gutter; *5* — foam bend sleeve; *6* — water bend sleeve;
8 — anode; *9* — cathode; *10* — tarp wall

This flotomachine action principle is analogous to flotation machine action principle presented on Fig. 3.1. Nevertheless, oxygen bubbles, extracted on anode, in given case come into aeration zone where they form flotation complexes with pollution, and hydrogen bubbles, extracted on cathode, do not participate in the flotation process.

In the given case, the process proceeds by the scheme presented on Fig. 3.5.

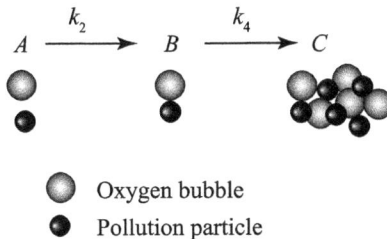

Fig. 3.5. Scheme of electro-flotation process on oxygen

In this scheme, A — pollution particles in initial state, B — flotocomplexes with oxygen bubbles, C — state of particles in foam layer;

constant k_2 characterizes flotocomplex formation probability while pollution particle interaction with oxygen bubble, and constant k_4 — surface emergence of given flotocomplex.

Differential equation system in this case will look in following way (3.15):

$$\begin{cases} \dfrac{dC_A}{dt} = -k_2 C_A; \\[2mm] \dfrac{dC_B}{dt} = k_2 C_A - k_4 C_B; \\[2mm] \dfrac{dC_C}{dt} = k_4 C_B. \end{cases} \qquad (3.15)$$

Initial conditions for given differential equation system at $t = 0$ are the following (3.16):

$$\begin{aligned} C_A(0) &= C_{A_0}; \\ C_B(0) &= 0; \\ C_C(0) &= 0. \end{aligned} \qquad (3.16)$$

This system solution has been obtained in the following form (3.17).

$$C_A(t) = C_{A_0} e^{-k_2 t};$$

$$C_B(t) = \frac{C_{A_0} k_1}{k_3 - k_1}\left(e^{-k_2 t} - e^{-k_4 t}\right);$$

$$C_C(t) = C_{A_0}\left(1 - \frac{k_2}{k_4 - k_2}e^{-k_2 t} - \frac{k_2}{k_2 - k_4}e^{-k_4 t}\right). \qquad (3.17)$$

Constant k_2, k_4 calculation is pursued analogously to the previous case according to (3.14) и (3.22) formulas correspondingly, oxygen electrochemical equivalent $k_e \approx 8{,}29 \cdot 10^{-8}$ kg/C; oxygen density $\rho = 1.47$ kg/m^3.

$$k_2 = \frac{1.5 \cdot 8.29 \cdot 10^{-8} \cdot 100 \cdot 0.05}{1 \cdot 50 \cdot 10^{-6} \cdot 1.47} = 0.00846\,\frac{1}{s};$$

$$k_4 = \frac{0.00136}{1} = 0.00136\,\frac{1}{s}.$$

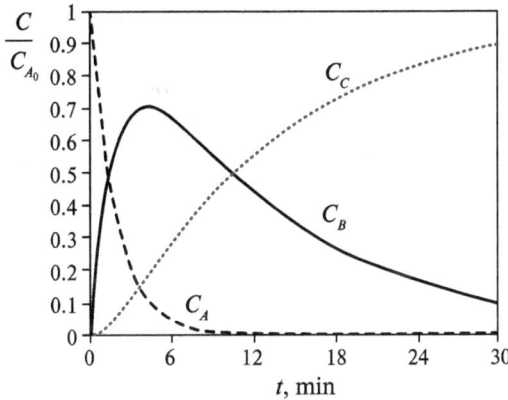

Fig. 3.6. Dependence of concentration from process time while electro-flotation process on oxygen

Dependence graphs of concentration from time with the use of obtained constants according to found solutions has been built (Fig. 3.6).

3.4 ELECTRO-FLOTATION ON HYDROGEN AND OXYGEN

The scheme of flotation machine, where electro-flotation process proceeds on the hydrogen and oxygen, is presented on Fig. 3.7.

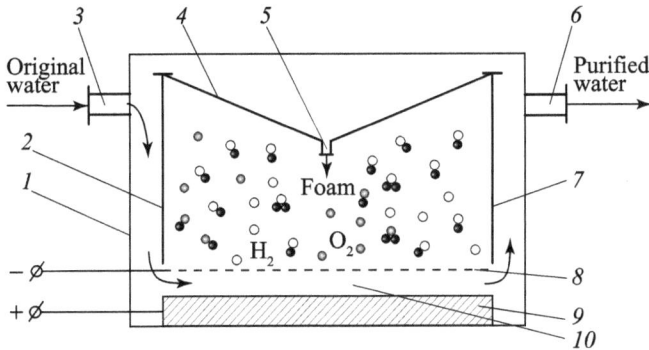

Fig. 3.7. Scheme of the flotation machine:
electro-flotation on hydrogen and oxygen:
1 — shell; *2,7* — semi-submersible walls, *3* — water supply sleeve;
4 — foam gutter; *5* — foam bend sleeve; *6* — water bend sleeve;
8 — cathode; *9* — anode

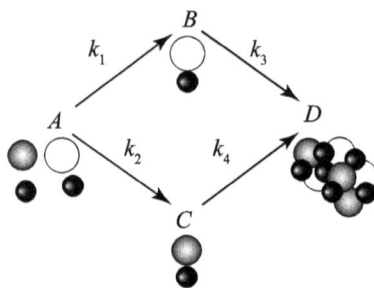

Fig. 3.8. Scheme of electro-flotation process on oxygen

○ Hydrogen bubble
◑ Oxygen bubble
● Pollution particle

This flotomachine action principle is analogous to flotation machine action principle presented on Fig. 3.1 and 3.4. Nevertheless, in this case, the wall between electrodes is absent and bubbles of both, hydrogen and oxygen, get into aeration zone.

The process goes on in this case according to the scheme presented on Fig. 3.8.

In the given scheme: A — pollution particles in initial state, B — flotocomplexes with hydrogen bubbles, C — flotocomplexes with oxygen bubbles; D — particle state in foam layer; constant k_1 characterizes flotocomplex formation probability while pollution particle interaction with hydrogen bubble, k_2 characterizes flotocomplex formation probability while pollution particle interaction with oxygen bubble, k_3 characterizes surface emergence of flotocomplex with hydrogen bubble, k_4 characterizes surface emergence of flotocomplex with oxygen bubble.

Differential equation system in this case will look in the following way (3.18):

$$
\begin{cases}
\dfrac{dC_A}{dt} = -k_1 C_A - k_2 C_A \\[2mm]
\dfrac{dC_B}{dt} = k_1 C_A - k_3 C_B \\[2mm]
\dfrac{dC_C}{dt} = k_2 C_A - k_4 C_C \\[2mm]
\dfrac{dC_D}{dt} = k_3 C_B + k_4 C_C
\end{cases}
\tag{3.18}
$$

Initial conditions for this equation system are the following at $t = 0$ (3.19):

$$C_A(0) = C_{A_0};$$
$$C_B(0) = 0;$$
$$C_C(0) = 0;$$
$$C_D(0) = 0.$$

(3.19)

Solution of the equation system can be found in analytical form. We have obtained given system solution of kind (3.19):

$$C_A(t) = C_{A_0} e^{-\alpha t};$$

$$C_B(t) = \frac{C_{A_0} k_1}{k_3 - \alpha} \left(e^{-\alpha t} - e^{-k_3 t} \right);$$

$$C_C(t) = \frac{C_{A_0} k_2}{k_4 - \alpha} \left(e^{-\alpha t} - e^{-k_4 t} \right);$$

(3.20)

$$C_D(t) = C_{A_0} \left\{ 1 + \frac{k_1 k_3}{k_3 - \alpha} \left(-\frac{1}{\alpha} e^{-\alpha t} + \frac{1}{k_3} e^{-k_3 t} \right) + \frac{k_2 k_4}{k_4 - \alpha} \left(-\frac{1}{\alpha} e^{-\alpha t} + \frac{1}{k_4} e^{-k_4 t} \right) \right\},$$

where $\alpha = k_1 + k_2$.

Graphs of concentration change versus time with the use of values of the constants according to found solution have been built (Fig. 3.9).

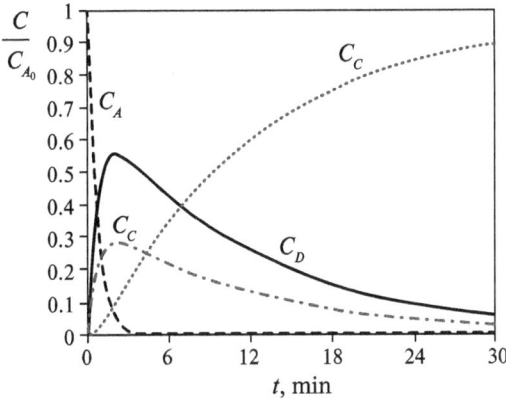

Fig. 3.9. Pollution concentration dependence from flotation time during pursuing of electro-flotation on hydrogen and oxygen

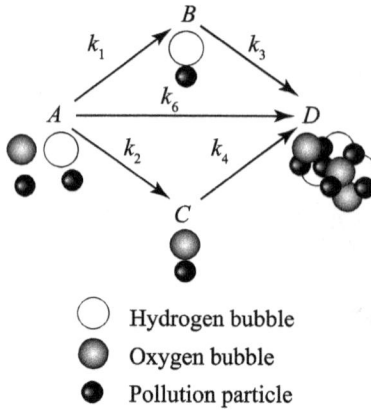

○ Hydrogen bubble
◉ Oxygen bubble
● Pollution particle

Fig. 3.10. Scheme of electro-flotation process of mineral oils

Data, presented on Fig. 3.3, 3.6, 3.9, confirm earlier obtained results of experimental research and give basis recommending considered approach for usage in calculation practice of electro-flotation purification of waste-waters.

It is needed for more precise description of considered process to take into account the constant characterizing autonomous surface emerging of oil drops (k_6). Process scheme in this case will have the following form (Fig. 3.10).

Differential equation system in this case will look in the following way (3.21):

$$
\begin{cases}
\dfrac{dC_A}{dt} = -k_1 C_A - k_2 C_A - k_6 C_A; \\[2mm]
\dfrac{dC_B}{dt} = k_1 C_A - k_3 C_B; \\[2mm]
\dfrac{dC_C}{dt} = k_2 C_A - k_4 C_C; \\[2mm]
\dfrac{dC_D}{dt} = k_3 C_B + k_4 C_C + k_6 C_A.
\end{cases}
\tag{3.21}
$$

Initial conditions for given system are the following at $t = 0$:

$$C_A(0) = C_{A_0};$$
$$C_B(0) = 0;$$
$$C_C(0) = 0;$$
$$C_D(0) = 0.$$

The comparison of effectiveness and duration of electro-flotation process has been pursued on the basis of author own results and those obtained together with colleagues as well as on the known experimental data [31]. According to the data, elctroflotation effectiveness and process duration depend on many parameters: extracted pollution kind and initial concentration, electrode kind, reagent presence, current density, pH and so on (see Table 3.1).

Table 3.1. Experimental data of OJSC "GosNIIsynthesbelok"
(Ksenofontov B.S. 2010)

Electrode kind	Substance being extracted	№	Current density mA/sm²	Reagent	Concentration, mg/l		Efficiency, %	Time, min
					before	after		
Cathode: steel 3	p. *Pseudo-monas* bacteria cells	1	5	without reagent	300	63.9	78.7	25
Anode: graphite		2	10	—«—	300	32.7	89,1	25
		3	15	—«—	300	21.6	92.8	25
		4	20	—«—	300	29.1	90.3	25
	Iron	1	5	Active silt (1 g/l)	15	1.89	87.4	30
		2	10	Active silt (5 g/l)	15	0.81	94.6	30
		3	15	(10 g/l)	15	1.17	92.2	30
		4	20	Active silt (50 g/l)	15	1.56	89.6	30

Table 3.2. Electro-flotation process efficiency
on the basis of [24] reference data

Electrode kind	Substance being extracted	№	Current density mA/sm²	Reagent	Concentration, mg/l		Efficiency, %	Time, min
					before	after		
Cathode: aluminium	bentonite	1	46.6	NaCl (1 g/l)	500	95	81	15
		2	48.3	NaCl (5 g/l)	500	46.5	90.7	15
Anode: graphitic		3	70	NaCl (10 g/l)	500	35	82.7	15
		4	198.3	NaCl (50 g/l)	500	35	80	15
	mineral oils	1	46.6	NaCl (1 g/l)	1000	260	74	15
		2	48.3	NaCl (5 g/l)	1000	130	87	15
		3	70	NaCl (10 g/l)	1000	69	93.1	15
		4	198.3	NaCl (50 g/l)	1000	93	90.7	15

Submitted data analysis shows that the fullest extraction is reached after about 25–30 minutes.

Let's consider process of water purification from suspended substances and mineral oils on the example of work [24] results. Experimental data on efficiency definition of oil industry wastewater electro-flotation are presented in [24]. Process major characteristics and electro-flotation process efficiency are presented in Table 3.2.

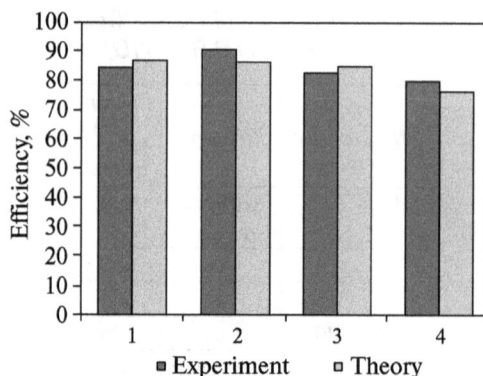

Fig. 3.11. Comparison of theoretical and experimental data
(on bentonite extraction)

Submitted data in Table 3.2 point out electro-flotation approach high efficiency while extraction, for example, of concrete and mineral oils.

The comparison of data obtained during theoretical calculation of results with experimental data, received in work [24], are shown in Table 3.3 and on Fig. 3.11.

The comparison of data obtained while theoretical calculation with experimental data received in work [24], are shown in Table 3.4 and on Fig. 3.12.

Table 3.3. Comparison of theoretical calculation with experimental data at flotation time t = 15 min (on bentonite extraction)

№	Current density, mA/sm²	Efficiency, %		Bial, %
		Experiment	Theory	
1	46.6	85.0	86.6	1.8
2	48.3	90.7	86.5	4.6
3	70	82.7	85.0	2.7
4	19.3	80.0	76.3	4.6

Table 3.4. Comparison of theoretical and experimental data at flotation time t = 15 min (on mineral oils extraction)

№	Current density, mA/sm²	Efficiency, %		Bial, %
		Experiment	Theory	
1	46.6	74.0	83,8	13.24
2	48.3	87.0	83	4.5
3	70	93.1	82.8	11
4	19.3	90.7	76.8	15.3

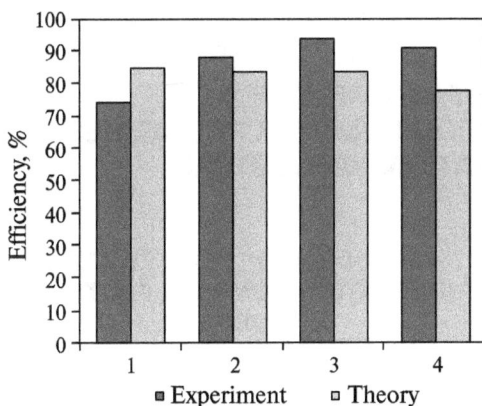

Fig. 3.12. Comparison of theoretical and experimental data on the example of mineral oils extraction from wastewaters

Above-presented data point out on good convergence of theoretical and experimental data. It's worth to pinpoint in a whole that electro-flotation approach, as our experience shows and other author's evidences confirm [22, 23], can be used for purification of wastewaters of various content.

4 MODELS OF WASTEWATER PURIFICATION COMPLEX PROCESSES AND FLOTATION HARVESTERS

4.1 MODELS OF WASTEWATER PURIFICATION COMPLEX FLOTATION PROCESSES

Flotation purification is rather widely used in practical technologies of water processing. Its intensification is much determined by the use of models, reflecting the reality of processes, proceeding during wastewater purification [50]. In this context, let's consider the complexes, for example, during wastewater flotopurification from metals. Fig. 4.1 shows different variants of formation of such complexes, which can occur, for instance, during wastewater flotopurification from metals.

The interaction of complexes between each other and their transfer from one state to another constitute the basis for models of processes of wastewater physical-chemical purification.

In consequent processes a product of one stage passes into next stage.

Consequent processes can contain from two to several thousand stages.

For example, simple case of multistage flotation model.

$$A \xrightarrow{k_1} B \xrightarrow{k_2} CA \xrightarrow{k_1} B.$$

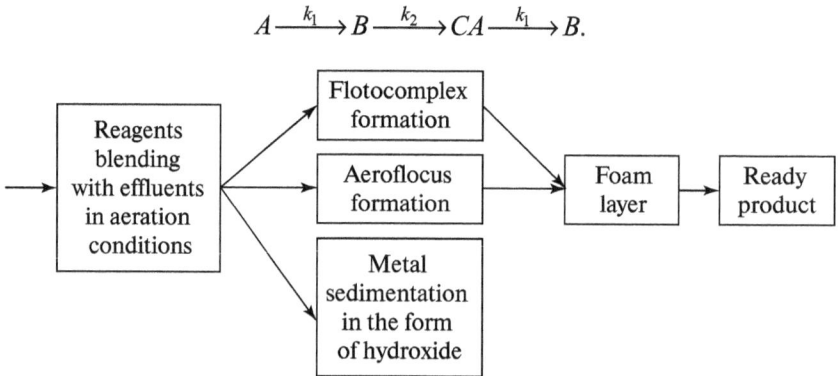

Fig. 4.1. Variants of formation of various complexes during flotopurificion of wastewaters from metals

For initial conditions $t = 0$, $C_A = a_0$, $C_B = C_C = 0$, kinetic equations will be the following:

$$\begin{cases} \dfrac{dC_A}{dt} = -k_1 C_A; \\ \dfrac{dC_B}{dt} = k_1 C_A - k_2 C_B. \end{cases}$$

Solution:

$$C_A = a_0 e^{-k_1 t};$$

$$C_B = \frac{C_A k_1}{k_2 - k_1}\left(e^{-k_1 t} - e^{-k_2 t}\right).$$

At

$$t = t_m = \frac{1}{k_2 - k_1}\ln\frac{k_2}{k_1},$$

intermediate product concentration reaches the maximum:

$$C_{B_{max}} = a_0 \left(\frac{k_2}{k_1}\right)^{\frac{k_2}{k_1 - k_2}}.$$

End product concentration:

$$C_C = a_0\left(1 - \frac{k_2}{k_2 - k_1}e^{-k_1 t} + \frac{k_1}{k_2 - k_1}e^{-k_2 t}\right).$$

Let's consider the case of model of mega-aerofloccules in the process of active silt flotation:

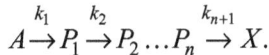

$$A \xrightarrow{k_1} P_1 \xrightarrow{k_2} P_2 \ldots P_n \xrightarrow{k_{n+1}} X.$$

For initial conditions:

$$t = 0,$$

$$C_A = a_0,$$

$$C_{P_1} = C_{P_2} = \ldots C_{P_n} = 0,$$

the solution for intermediate products will be the following:

$$C_{P_i} = a_0 \prod_{j=1}^{i} k_j \sum_{j=1}^{i+1} \frac{e^{-k_j t}}{\prod_{m=1}^{i+1}(k_m - k_j)},$$

where $m = j$.

It represents an interest a case of wastewater conditioning with the use of reagent — collector B, interacting with original pollution A. Whereat, P — state of flotocomplex "particle — gas bubble", and D — state in a foam layer.

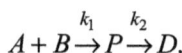

$$A + B \xrightarrow{k_1} P \xrightarrow{k_2} D.$$

For initial conditions $t = 0$, $CA = a_0$, $CB = b_0$, $CP = CD = c_0$, the kinetic equations will be the following:

$$\begin{cases} \dfrac{dC_A}{dt} = \dfrac{dC_B}{dt} - k_1 C_A C_B; \\ \dfrac{dC_P}{dt} = k_1 C_A C_B - k C_P. \end{cases}$$

Solution:

$$C_A = a_0 \frac{b_0 - a_0}{b_0 e^{-k_1(b_0 - a_0)t} - a_0};$$

$$C_B = b_0 - (a_0 - C_A);$$

$$C_P = -\frac{a_0}{1 + a_0 k_1 t} + b_0 e^{-k_2 t} + \frac{k_2}{k_1} e^{\left(-k_1 k_2 a_0^{-1} + k_2 t\right)} \left(l_i e^{\left(k_1^{-1} k_2 a_0^{-1} + k_2 t\right)} - l_i e^{k_1^{-1} k_2 a_0^{-1}} \right).$$

where $l_i x = \int_0^x \dfrac{dz}{\ln z}$ — integral logarithm.

The cases of microflotocomplexes formation and their coalescence according to the scheme are rather interesting:

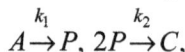

$$A \xrightarrow{k_1} P, \; 2P \xrightarrow{k_2} C,$$

where A — original pollution;

P — non-floating microflotocomplex (flotocomplex which is late to emerge on surface into foam layer while being in flotation zone because of small lifting force of a microbubble);

$2P$ — floating microflotocomplex (flotocomplex which can emerge on surface into foam layer in flotation zone at longtime being in the zone).

For initial conditions $t = 0$, $CA = a_0$, $CP = CC = 0$, the solution will be the following:

$$C_P = a_0 \left(\frac{e^{-xt}}{x}\right)^{\frac{1}{2}} \cdot \frac{iJ_1\left[2i\left(xe^{-xt}\right)^{\frac{1}{2}}\right]}{J_0\left[2i\left(xe^{-xt}\right)^{\frac{1}{2}}\right]} + \frac{\beta H_1^{(1)}\left[2i\left(xe^{-xt}\right)^{\frac{1}{2}}\right]}{\beta H_0^{(1)}\left[2i\left(xe^{-xt}\right)^{\frac{1}{2}}\right]},$$

where $x = a_0 \dfrac{k_2}{k_1}$;

$$\beta = \frac{iJ_1\left(2i\sqrt{x}\right)}{H_1^{(1)}\left(2i\sqrt{x}\right)};$$

i — imaginary unit;

J_0, J_1 — Bessel functions;

$H_0^{(1)}, H_1^{(1)}$ — Gangle functions.

Let's consider floating of pollutions in the form of microflotocomplexes according to the scheme:

$$2A \xrightarrow{k_1} P \xrightarrow{k_2} C,$$

where A — original pollution in the form of non-floating microflotocomplex;

P — floating microflotocomplex; C — state in foam layer.

For initial conditions $t = 0$, $CA = a_0$, $CP = CC = 0$, the solution will be the following:

$$C_P = \frac{1}{2}a_0\left\{e^{-\eta(\tau-1)} + \eta e^{-\eta\tau}E_i\left(\eta\tau\right) - E_i\left(\eta\right)\tau^{-1}\right\},$$

where $\eta = \dfrac{k_2 a_0}{k_1}$;

$\tau = 1 + a_0 k_1 t$;

$$C_{P_{max}} = \frac{1}{2}\eta\tau_{max}^2 a_0;$$

E_i —integral exponential function.

Let's consider the floating of sub-microflotocomplexes according to the scheme:

$$2A \xrightarrow{k_1} P, \ 2P \xrightarrow{k_2} C,$$

where A — initial pollution in the form of non-floating sub-microfloto-complex (flotocomplex which cannot surface-emerge into foam layer during whatever longtime being in flotation zone because of very small lifting force of a microbubble);
P — non-floating microflotocomplex;
C — state in foam layer.

For initial conditions $t = 0$, $CA = a_0$, $CP = CC = 0$, the solution will be the following:

$$C_P = \frac{a_0}{2\eta\tau}\left[(\alpha+1) - 2\alpha\left(1 + \frac{\alpha-1}{\alpha+1}\tau\alpha\right)^{-1}\right];$$

$$C_{P\max} = a_0\tau_{\max}\left(\frac{1}{2\eta}\right)^{\frac{1}{2}},$$

where $\eta = \dfrac{k_1}{k_2}$;

$\tau = 1 + a_0 k_1 t$;

$\alpha = (1+2\eta)^{\frac{1}{2}}$.

The consideration of processes with limiting stage is of big practical interest.

If the process includes a series of consequent stages, and speed constant of one of them is much less than speed constant of other stages, this stage is a limiting one, namely, it defines the speed of all of the process. Said is characteristic, for instance, for purification processes with the use of ionic flotation, proceeding according to the scheme:

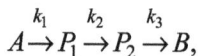

$$A \xrightarrow{k_1} P_1 \xrightarrow{k_2} P_2 \xrightarrow{k_3} B,$$

where A — original pollution (metal ion);
B — complex "metal ion — collector";
C — flotocomplex "metal ion — collector — gas bubble";
X — state in foam layer.

42

If $k_1 < k_2$ and $k_1 < k_3$, the limiting stage is the first one, and if $k_2 < k_1$ and $k_2 < k_3$, the limiting stage is the second one and so on. When the orders of consequent stages are different, not speed constants but specific speeds of consequent stages should be compared. For instance, in the case of the process proceeding by the scheme:

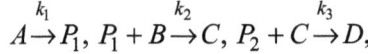

$$A \xrightarrow{k_1} P_1, \ P_1 + B \xrightarrow{k_2} C, \ P_2 + C \xrightarrow{k_3} D,$$

where the second stage will limit in the condition $k_2 CB < k_1$ и $k_2 CB < k_3 CC$.

The scheme of consequent-parallel processes of floating and sedimentation:

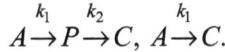

$$A \xrightarrow{k_1} P \xrightarrow{k_2} C, \ A \xrightarrow{k_1} C.$$

For initial conditions $t = 0$, $CA = a_0$, $CP + CC = 0$, kinetic equations:

$$\begin{cases} \dfrac{dC_A}{dt} = -(k_1 + k_1')C_A; \\[2mm] \dfrac{dC_P}{dt} = k_1 C_A - k_2 C_P; \\[2mm] \dfrac{dC_C}{dt} = k_1' C_A + k_2 C_P. \end{cases}$$

Solutions:

$$C_P = \frac{a_0 k_1}{k_2 - k_1 - k_1'}\left[e^{-(k_1 + k_1')t} - e^{-k_2 t} \right],$$

$$C_P = \frac{a_0 k_1'}{k_1 + k_1'}\left(1 - e^{-(k_1 + k_1')t}\right) + \frac{a_0 k_1}{k_2 - k_1 - k_1'}\left\{ 1 - e^{-k_2 t} + \frac{k_2}{k_1 + k_1'}\left(1 - e^{-(k_1 + k_1')t}\right) \right\},$$

$$\frac{dC_C}{d\Delta C_A} = \frac{k_1}{k_1 + k_1'} + \frac{k_2}{k_1 + k_1'}\frac{C_p}{C_A}.$$

Let consider the process of floccule formation with the use of coagulants and flocculants which goes on according to the scheme:

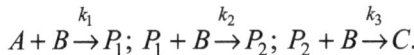

$$A + B \xrightarrow{k_1} P_1; \ P_1 + B \xrightarrow{k_2} P_2; \ P_2 + B \xrightarrow{k_3} C.$$

Initial pieces of kinetic curves are described by the equations:

$$\begin{cases} \dfrac{dC_{P_1}}{dt} = k_1 C_A C_B, \quad C_{P_1} \cong k_1 a_0 b_0 t; \\[2mm] \dfrac{dC_{P_2}}{dt} = k_2 C_A C_{P_1}, \quad C_{P_2} \cong \dfrac{1}{2} k_1 k_2 a_0 b_0^2 t^2; \\[2mm] \dfrac{dC_C}{dt} = k_3 C_B C_{P_2}, \quad C_{P_2} \cong \dfrac{1}{6} k_1 k_2 k_3 a_0 b_0^3 t^3, \end{cases}$$

and ratios between concentrations by the equations:

$$-\frac{dC_{P_1}}{dC_A} = \frac{k_1 C_A C_B - k_2 C_B C_{P_1}}{k_1 C_A C_B} = \frac{\eta_1 C_A - C_{P_1}}{\eta_1 C_A},$$

$$-\frac{dC_{P_2}}{dC_A} = \frac{k_2 C_B C_{P_1} - k_3 C_B C_{P_2}}{k_1 C_A C_B} = \frac{\eta_1 C_{P_1} - C_{P_2}}{\eta_1 \eta_2 C_A}.$$

Solutions:

$$C_{P_1} = \frac{\eta_1}{1-\eta_1} C_A + \frac{\eta_1}{\eta_1 - 1}\left(\frac{C_A}{a_0}\right)^{\frac{1}{\eta_1}} a_0;$$

$$\frac{C_{P_2}}{a_0} = \frac{\eta_1 \eta_2}{(1-\eta_2)(\eta_1 - 1)}\left(\frac{C_A}{a_0}\right)^{\frac{1}{\eta_1}} + \frac{\eta_1 \eta_2 C_A}{a_0(1-\eta_1)(1-\eta_1\eta_2)} +$$

$$+ \frac{\eta_1 \eta_2^2}{(1-\eta_1)(1-\eta_1\eta_2)}\left(\frac{C_A}{a_0}\right)^{\frac{1}{\eta_1\eta_2}},$$

where $\eta_1 = \dfrac{k_1}{k_2}, \eta_2 = \dfrac{k_2}{k_3}$.

In the row of cases, microflotocomplex floating can be described by the scheme:

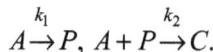

$$A \xrightarrow{k_1} P, \; A + P \xrightarrow{k_2} C.$$

For initial conditions $t = 0$, $C_A = a_0$, $C_P = C_C = 0$, kinetic equations will be the following:

44

$$\begin{cases} \dfrac{dC_A}{dt} = -(k_1 C_A + k_2 C_A C_P); \\ \dfrac{dC_P}{dt} = k_1 C_A - k_2 C_A C_P. \end{cases}$$

Solutions:

$$C_P + 2\eta \ln \frac{\eta - C_P}{\eta} = C_A - a_0;$$

$$\frac{C_P}{\eta} + 2\ln\left(1 - \frac{C_P}{\eta}\right) = -\frac{a_0}{\eta}\left(1 - \frac{C_A}{a_0}\right).$$

Stationary value of intermediate product concentration:

$$C_{P_{stat}} = \eta = \frac{k_1}{k_2}.$$

In the case of floating of microflotocomplexes with reagents (coagulants), we have:

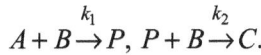

$$A + B \xrightarrow{k_1} P, \ P + B \xrightarrow{k_2} C.$$

For initial conditions $t = 0$, $CA = a_0$, $CB = b_0$, $CP + CC = 0$, kinetic equations look as the following:

$$\begin{cases} \dfrac{dC_A}{dt} = -k_1 C_A C_B; \\ \dfrac{dC_B}{dt} = -(k_1 C_A C_B + k_2 C_B C_P); \\ \dfrac{dC_P}{dt} = k_1 C_A C_B - k_2 C_B C_P. \end{cases}$$

Solution:

$$\frac{C_P}{C_A} = \frac{1}{\eta - 1}\left[1 - \left(\frac{C_A}{a_0}\right)^{n-1}\right],$$

where $\eta = \dfrac{k_2}{k_1}$.

Intermediate product maximal concentration:

$$C_{P_{max}} = a_0 \eta^\varepsilon,$$

where $\varepsilon = \dfrac{\eta}{1-\eta}$ for $h < 1,$;

$$C_{P_{max}} \dfrac{a_0}{e} \text{ for } h = 1.$$

Let's consider the process of coalescence and floating of microfloto-complexes, described by the scheme:

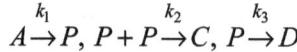

$$A \xrightarrow{k_1} P, \; P + P \xrightarrow{k_2} C, \; P \xrightarrow{k_3} D$$

For initial conditions $t = 0$, $CA = a_0$, $CP = CC = CD = 0$, kinetic equations will be the following:

$$\begin{cases} \dfrac{dC_A}{dt} = -k_1 C_A; \\ \dfrac{dC_P}{dt} = k_1 C_A - k_2 C_P{}^2 - k_2 C_P). \end{cases}$$

Solutions:

if $\dfrac{k_3}{k_1}$ — an integer,

$$C_P = \frac{k_1\sqrt{\eta_2}}{k_2} \cdot \frac{N_{-\eta_1}\left(2\sqrt{\eta_2}\right)J_{-\eta_1}\left(2\sqrt{\eta_2}\sigma\right) - J_{-\eta_1}\left(2\sqrt{\eta_2}\right)N_{-\eta_1}\left(2\sqrt{\eta_2}\sigma\right)}{N_{-\eta_1}\left(2\sqrt{\eta_2}\right)J_{1-\eta_1}\left(2\sqrt{\eta_2}\sigma\right) - J_{-\eta_1}\left(2\sqrt{\eta_2}\right)N_{1-\eta_1}\left(2\sqrt{\eta_2}\sigma\right)};$$

$$\eta_1 = 1 - \frac{k_3}{k_1}, \; \eta_2 = a_0\frac{k_2}{k_1}, \; \sigma = e^{-\frac{1}{2}k_1 t},$$

if $\dfrac{k_3}{k_1}$ — not an integer,

$$C_P = \frac{k_1\sqrt{\eta_2}}{k_2} \cdot \frac{J_{\eta_1}\left(2\sqrt{\eta_2}\right)J_{-\eta_1}\left(2\sqrt{\eta_2}\sigma\right) - J_{-\eta_1}\left(2\sqrt{\eta_2}\right)J_{\eta_1}\left(2\sqrt{\eta_2}\sigma\right)}{J_{\eta_1}\left(2\sqrt{\eta_2}\right)J_{1-\eta_1}\left(2\sqrt{\eta_2}\sigma\right) - J_{-\eta_1}\left(2\sqrt{\eta_2}\right)J_{1-\eta_1}\left(2\sqrt{\eta_2}\sigma\right)},$$

where J — Bessel functions;
N — Neumann functions.

Let consider consequent processes with equilibrium stages.

For instance, the floating with reversible first stage is described by the scheme:

$$A \overset{k_1, \; k_2}{\longleftrightarrow} B \overset{k_3}{\rightarrow} C.$$

For initial conditions $t = 0$, $CA = a_0$, $CP = CC = CD = 0$, kinetic equations look the following way:

$$\begin{cases} \dfrac{dC_A}{dt} = -k_1 C_A + k_2 C_B; \\[2mm] \dfrac{dC_B}{dt} = k_1 C_A - (k_2 + k_3)C_B; \\[2mm] \dfrac{dC_C}{dt} = k_3 C_B. \end{cases}$$

Solutions:

$$C_A = a_0 \left\{ \dfrac{\lambda_2 - k_1^{-1}k_3}{\lambda_2(\lambda_2 - \lambda_3)} e^{-\lambda_2 k_1 t} + \dfrac{k_1^{-1}k_3 - \lambda_3}{\lambda_3(\lambda_2 - \lambda_3)} e^{-\lambda_3 k_1 t} \right\},$$

$$C_B = \dfrac{a_0}{\lambda_2 - \lambda_3} \left\{ e^{-\lambda_3 k_1 t} - e^{\lambda_2 k_1 t} \right\},$$

where

$$\lambda_2 = \dfrac{1}{2}(\alpha + \beta), \quad \alpha = 1 + \dfrac{k_2}{k_1} + \dfrac{k_3}{k_1}, \quad \lambda_3 = \dfrac{1}{2}(\alpha - \beta), \quad \beta = \left(\alpha^2 - \dfrac{4k_3}{k_1} \right)^{\frac{1}{2}}.$$

The floating with reversible second stage is described by the scheme:

$$A \overset{k_1}{\rightarrow} B \overset{k_2, \; k_3}{\longleftrightarrow} C.$$

For initial conditions $t = 0$, $CA = a_0$, $CB = CC = 0$, $t \rightarrow \infty$, $CB \rightarrow CB$, $CC \rightarrow CC$, the kinetic equations will look the following way:

$$\begin{cases} \dfrac{dC_A}{dt} = -k_1 C_A; \\[2mm] \dfrac{dC_B}{dt} = k_1 C_A + k_3 C_C - k_2 C_B; \\[2mm] C_A = a_0 e^{-k_1 t}. \end{cases}$$

Solutions:

$$C_B = \frac{C_{B\infty}}{k_1 + k_2 - k_3}\left\{a_0\left(k_1 - k_2\right)e^{-k_1 t} - \left[k_2 C_{B\infty} - C_{C\infty}\left(k_3 - k_1\right)\right]e^{-(k_2 + k_3)t}\right\};$$

$$C_C = \frac{C_{C\infty}}{-(k_1 + k_2 - k_3)}\left\{a_0 k_2 e^{-k_1 t} - \left[k_2 C_{B\infty} - C_{C\infty}\left(k_3 - k_1\right)\right]e^{-(k_2 + k_3)t}\right\}.$$

The floating with collector and flocculant is going on by the scheme:

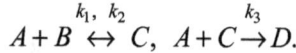

$$A + B \overset{k_1,\ k_2}{\leftrightarrow} C,\ A + C \overset{k_3}{\to} D.$$

For initial conditions $t = 0$, $CA = a_0$, $CB = b_0$, $CC = CD = 0$, kinetic equations will be the following:

$$\begin{cases} \dfrac{dC_A}{dt} = -k_1 C_A C_B + k_2 C_C - k_3 C_A C_C; \\[2mm] \dfrac{dC_B}{dt} = -k_1 C_A C_B + k_2 C_C; \\[2mm] \dfrac{dC_P}{dt} = k_1 C_A C_B - k_2 C_C - k_3 C_A C_C. \end{cases}$$

If at $t = 0$, $CC = c_0$, the concentrations of the rest of products are equal to zero, and whereat, the equilibrium concentration is being set

$$C_A = \frac{C_C k_2}{C_B k_1},$$

we have

$$\frac{dC_C}{dt} = \frac{2C_C^2 k_2 k_3}{C_B k_1} \quad \text{and} \quad \frac{dC_C}{W_C} = \frac{k_1}{2k_2 k_3} \cdot \frac{C_B}{C_C}.$$

The floating with the use of various reagents is described by the scheme:

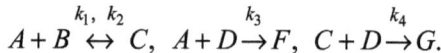

$$A + B \overset{k_1,\ k_2}{\leftrightarrow} C,\ A + D \overset{k_3}{\to} F,\ C + D \overset{k_4}{\to} G.$$

Let's suppose that stationary concentration C is being quickly reached in a system, then

$$C_C = \frac{k_1 C_A C_B}{k_2 + k_4 C_A};$$

$$W = \frac{dC_A}{dt} = \frac{k_1 k_4 C_A C_B C_D}{k_2 + k_4 C_A} + k_3 C_A C_D;$$

$$k_{\exp} = \frac{W}{C_A} = \frac{\Delta \ln C_A}{\Delta t} = \left(\frac{k_1 k_4 C_B}{k_2 + k_4 C_D} + k_3 \right) C_D = k' C_D;$$

$$\lim_{C_A \to 0} k' = k_3, \quad \lim_{C_D \to 0} k' = \frac{k_1 k_4 C_B}{k_2} + k_3, \quad \lim_{C_D \to \infty} \overline{k_{\exp}} = k_1 C_B + k_3 C_D.$$

Proposed solutions can be used in the practice of flotation thickening of active silt, particularly, at more precise calculation of flotation machines and apparatus with different aeration systems. Whereat, considered cases comprise the majority of processes of wastewaters flotation purification that is especially important during the use of combined flotation devices, including flotation harvesters [13], which are also applied in wastewater purification practice.

4.2 MODELLING OF PROCESSES IN FLOTATION HARVESTERS

Let's consider, first of all, major stages of processes, proceeding in flotation harvester's working space, during wastewater purification. Flotoharvester scheme, worked out by us [33], is presented on Fig. 4.2.

Flotoharvester for silt mixture division includes Shell *1*, which, exterior side on, are Sleeves *2* for silt mixture supply, defecated water Bend *3*, foam product Output *5*, Ejector *19* with flotomud output Sleeve *9*, Hydro-cyclone Collector *8* with output Sleeve *10* of thickened product and Bend Sleeves *7* of liquid phase, thickened flotomud bends of thickened flotomud *11* and of defecated liquid *6*, sediment output *12*, working liquid supply *21*, with Shell *13* of sediment, additionally installed on exterior side, this shell consists of internal Chamber *14* and external managing devices *15*, snap Collar *16* and Tray *17* for filtrate collection. Together, perforated Walls *20* and purified water regulation Device *18* are situated inside.

Work principle of flotoharvester is in the following. Original silt mixture comes into Shell *1* via Sleeve *2* of flotoharvester where its mixing with working liquid, entering through Sleeve *21*, proceeds. As a result of these flows mixing, there proceeds the formation of flotocomplexes (active silt) — (gas bubbles) which are present in working liquid. Formed flotocomplexes with

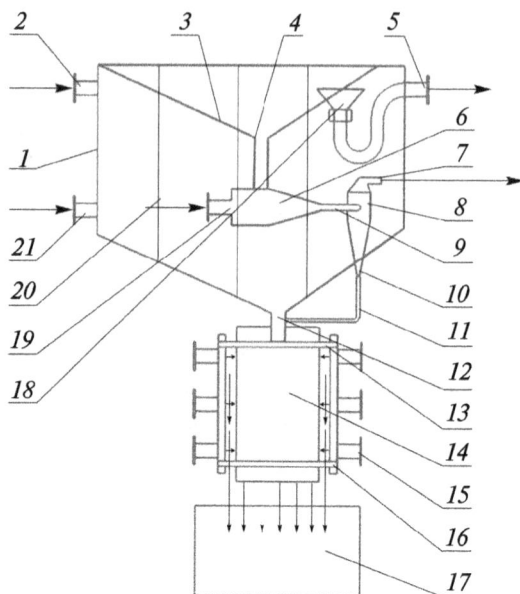

Fig. 4.2. Scheme of flotoharvester KBS-1
with block of sediment dehydration
(RF patent on useful model № 72180; author Ksenofontov B.S.)

large bubbles of about 1 mm and more emerge into foam layer rather quickly, and flotocomplexes with gas smaller size bubbles, called microflotocomplexes, are tended by the flow of being purified liquid which is filtered further via perforated Walls *20*. Microflotocomplexes coalesce while going through perforations; they unite in larger flotocomplexes and then emerge on surface into foam layer which is being gathered in foam Gutter *4* and then led out via Sleeve *5*. Flotomud after that is drawn into Ejector *19* where the destruction of foam product and its conversion into liquid, containing original pollutions, proceeds. Further, thickened concentrate comes via Pipeline *11* into interior Space *14* of Shell *13* wherein the bag from synthetic tissue is located. The dehydration of the sediment, being in the bag, goes on under the effect of external efforts from managing Devices *15*. Whereat, filtrate, which is being collected in Tray *17*, is drained, and dehydrated sediment is being removed together with the bag.

The clean water is being drawn from working space of Shell *1* with the help of Devices *18* and Sleeve *6*.

It's worth to note that with the use of flotoharvester with vibrating block, one can get rather high degree of active silt division from the water, exceeding on 20–25% in the row of cases of similar identifier with the use of well-known apparatus-analogs. Besides, in this case, the sediment, dehydrated till residual moisture content of 85–90%, is made.

Let's fulfill by stages the assessment of going on in flotoharvester processes.

After silt mixture and working liquid entering the harvester working space, flotocomplexes particle-bubble formation proceeds. Whereat, as a rule, not all particles of active silt merge with air particles and, being left in sole state or in the form of aggregates, fall out into sediment.

The particles, merged with small bubbles, form microflotocomplexes which slowly emerge on surface and, in this connection, the flow of purifying liquid, moving in horizontal direction, takes them. Such microflotocomplexes, reaching a net wall, contact between each other with the formation, as a rule, of larger bubbles which quickly emerge on surface, forming flotomud. Purified water, going through the net wall, bends with the help of special device and further goes out from flotoharvester via output sleeve. Water purification effect in this case significantly exceeds reached results in installations-analogs. Let's consider the schemes of processes in flotoharvestres of above-mentioned type.

Let's denote pollution concentration in water in the form of C where part of pollutions, differing by hydrophobic properties, is C_1, and another part in the form of C_2 are hydrophilic pollutions.

Whereat,

$$C = C_1 + C_2.$$

Let extraction flotation process goes on under the following scheme:

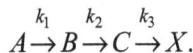

$$A \xrightarrow{k_1} B \xrightarrow{k_2} C \xrightarrow{k_3} X.$$

For initial conditions: $t = 0$, $C_1 A = a_0$, $C_1 = C_1 C = C_1 X = 0$, kinetic equations have the view:

$$\frac{dC_A^1}{dt} = -k_1 C_A^1;$$

$$\frac{dC_B^1}{dt} = k_1 C_A^1 - k_2 C_B^1;$$

$$\frac{dC_C^1}{dt} = k_2 C_B^1 - k_3 C_C^1.$$

Constants $k_1...k_3$ characterize speeds of transfer processes of extracted hydrophobic particles from state A into B, C and X with the obtaining of thickened flotomud.

Solution:

$$C_A^1 = a_0 e^{(-k_1 t)};$$

$$C_B = \frac{a_0 k_1}{k_2 - k_1} \left[e^{(-k_1 t)} - e^{(-k_2 t)} \right];$$

$$C_C^1 = k_1 k_2 a_0 \left(\frac{e^{(-k_1 t)}}{k_3 - k_1} + \frac{e^{(-k_2 t)}}{(k_2 - k_1)(k_2 - k_3)} + \frac{e^{(-k_3 t)}}{(k_3 - k_1)(k_3 - k_2)} \right);$$

$$C_X^1 = a_0 \left(1 - \frac{k_2 k_3 e^{(-k_1 t)}}{(k_2 - k_1)(k_3 - k_1)} - \frac{k_1 k_3 e^{(-k_2 t)}}{(k_2 - k_1)(k_2 - k_3)} - \frac{k_1 k_2 e^{(-k_3 t)}}{(k_3 - k_1)(k_3 - k_2)} \right).$$

For accompanying process of defecating, we have:

$$A \xrightarrow{k_4} D \xrightarrow{k_5} Y,$$

with initial conditions: $t = 0$, $C_2 A = b_0$, $C_2 D = C_2 Y = 0$, we can write the following kinetic equations:

$$\frac{dC_A^2}{dt} = -k_4 C_A^2;$$

$$\frac{dC_D^2}{dt} = k_4 C_A^2 - k_5 C_D^2.$$

Here, constants $k_4...k_5$ characterize rates of transfer processes of extracted hydrophilic particles from state A to D and Y with the obtaining of thickened sediment.

Solution:

$$C_A^2 = b_0 e^{(-k_4 t)};$$

$$C_D^2 = \frac{k_4 C_A^2}{k_5 - k_4} \left[e^{(-k_4 t)} - e^{(-k_5 t)} \right].$$

At

$$t = t_m = \frac{1}{k_5 - k_4} \ln \frac{k_5}{k_4},$$

intermediate product concentration reaches the maximum:

$$C_{B_{\max}}^2 = b_0 \left(\frac{k_5}{k_4} \right)^{\frac{k_5}{k_4 - k_5}}.$$

End product concentration:

$$C_Y^2 = b_0 \left(1 - \frac{k_5}{k_5 - k_4} e^{(-k_4 t)} + \frac{k_4}{k_5 - k_4} e^{(-k_5 t)} \right).$$

Comparison of theoretical and experimental data points on possibility of usage of offered mathematical models in practical calculations. Whereat, it should be noted that theoretical data exceed experimental results. It points on that some suppositions are simplified and do not take into account some effects, though, probably, they do not cause essential influence on purification efficiency as the discrepancy doesn't lie beyond 10%. Such discrepancy order doesn't impact vividly the calculations of major mattering sizes of flotoharvester.

Improved variant of above-described apparatus is a flotation harvester [34] which includes Shell 1 (Fig. 4.3), which, exterior side from, Sleeves 2 and 3 are located, correspondingly, for recirculating liquid and original (dirty) water supply, as well as foam Gutter 4 with foam product output Sleeve 5, clean water output Sleeve 7 with regulating Faucet 8, sediment output Sleeve 14, which Sheath 15 of Bag 16, having pores 17, is worn on. Whereat, Sleeve 5 is connected with copulative Element 10 with Ejector 11, which is united with Hose 12 and Bag 16, pursuing oscillations in vertical state with the help of Vibro-platform 18, which filtrate Collector 19 is situated under with output Sleeve 20, connected with the help of Pipeline 21 with Pump 22. After Pump 22 on Pipeline 21, there are located Sleeves 23 for coagulant input, pair of Magnets 24, Sleeve 25 of flocculant supply and Chamber 26 of flocculus formation.

Semi-submersible Wall 6 with low window with Net 9 and perforated Element 13, connected with the help of Pipeline 27 with Sleeve 2, are installed inside Shell 1.

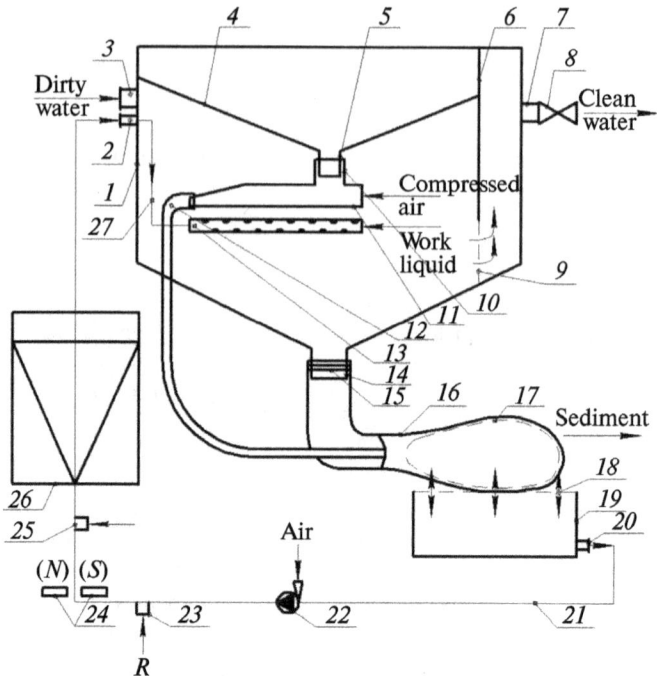

Fig. 4.3. Scheme of flotoharvester KBS-2
(RF patent № 2658411; author Ksenofontov B.S.):
1 — flotoharvester corps; *2* — sleeve for recirculating liquid supply;
3 — sleeve for polluted water supply; *4* — foam gutter; *5* — sleeve
of foam product output; *6* — semi-submersible wall; *7* — sleeve of
clean water bend; *8* — regulating faucet; *9* — filtrating net;
10 — connecting element; *11* — ejector; *12* — connecting hose;
13 — perforated element; *14* — sleeve of sediment bend;
15 — sheath; *16* — bag; *17* — pores; *18* — vibro-platform;
19 — filtrate collector; *20* — filtrate bend sleeve; *21* — pipeline; *22* —
pump; *23* — coagulant supply sleeve; *24* — pair of magnets; *25* —
sleeve of flocculant supply; *26* — conditioning chamber;
27 — pipeline of recirculating liquid supply

Work principle of flotoharvester is in the following. Dirty water, coming in via Sleeve *3*, mixes with recirculating and work liquids that leads to the formation of flotocomplexes and aggregates of pollution particles without bubbles which detach in about 15–30 minutes into foam layer and sediment, correspondingly. Foam layer is removed into foam Gutter *4* and then, via

54

Sleeve *5*, enter Ejector *11* by the way of aspiration and further along Hose *12* to Bag *16*. Entering Bag *16* the mixture of sediment and foam product is being dehydrated by the way of gravitational bend of moisture and simultaneous intensifying action of Vibro-platform *18*. The filtrate, formed whereat, is bent via Sleeve *19* and then is used as recirculating liquid, supplied into working space via Sleeve *2*. Dehydrated sediment is removed together with Bag *16*.

Clean water is drawn from work space of Shell *1* consequently through low window with Net *9* of semi-submersible Wall *6* and further via Sleeve *7* with the possibility of bent flow regulation.

The use of offered variant of flotoharvester allows to obtain active silt with the concentration of less than 10%.

It should be noted that flotation process's start in such apparatus proceeds on conditioning stage from forming of flotocomplexes particle-bubble. Whereat, as a rule, not all particles of pollutions stick to air bubbles and, staying in sole state or in the form of aggregates, fall out into sediment.

Particles, being coalescent with small bubbles, form microflotocomplexes which slowly emerge on surface and, in this connection, they are taken by the flow of purifying liquid moving in horizontal direction. Such microflotocomplexes, reaching the net wall, contact with each other with the formation, as a rule, of larger bubbles which quickly emerge on surface, forming flotomud. Purified water, going through the net wall, is bent with the help of special device from flotoharvester via exit sleeve. Water purification effect in this case significantly overtops the results reached on installations-analogs.

Other variant of flotoharvester [35] includes Shell *1* (Fig. 4.4) which, exterior side on, is Ejector *2* with sleeves for supply of silt mixture with Reagents *3*, Air *4*, foam Gutter *5* with exit Sleeve *7*, connected by Hose *11* with Sleeve *16*, and Sleeves of defecated water bending *10* and sediment *17*, and, inside corps, there are installed Walls *6* and wall with window, closed with Net *8* and Device *9* for defecated water level regulation. Sleeve *22* is installed for the supply of working liquid in the form of mixture water-air. Below Shell *1*, there is additionally located as a sediment thickening unit the screw Thickener *13* with installed inside of it Screw *15*, driven in rotation by Electro-engine *12*, and, from exterior side, there are Sleeve *14* for fugate output and Chamber *21* for thickened sediment collection with, installed inside it, false perforated Bottom *18* and with Sleeve *19* from exterior side for defecated water output and Sleeve *20* for thickened sediment output. Whereat, Screw *15* is accomplished with the length which is from 25:1 to 35:1 relatively screw diameter. Wall *8*, installed inside of flotoharvester corps, with window, closed with net

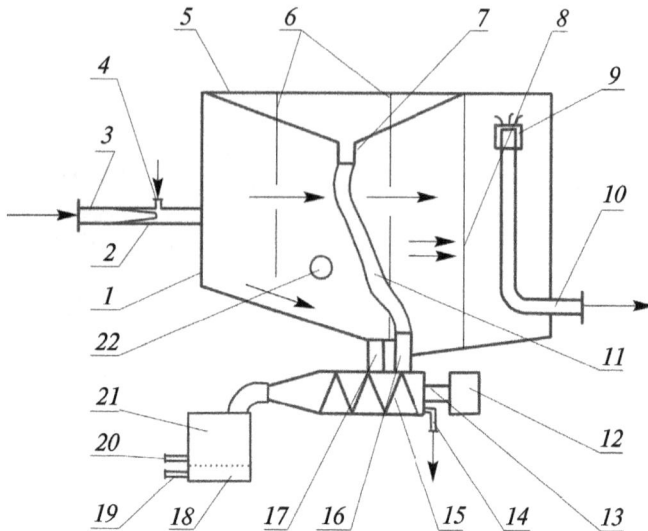

Fig. 4.4. Scheme of flotoharvester KBS-3
(RF patent on invention № 2669842; author Ksenofontov B.S.)

with cell size from 0.01 to 0.7 mm. Whereat, the net is accomplished with magnetic material, and sediment collecting chamber has perforated false bottom with live section from 20 to 60% from overall bottom area; together, false bottom perforations have the size of quadrates with quadrate side sizes from 0.1 to 0.9 mm.

Work principle of proposed flotoharvester proceeds in the following way. Silt mixture is given to the inside of Shell *1* of flotoharvester via Sleeve *3* of Ejector *2*. Meanwhile, air is supplied at the expense of pumping via Sleeve *4*. Then (silt mixture)-air comes inside Shell *1* where simultaneously, via Sleeve *22*, the working liquid in the form of purified water with diluted air therein is supplied. At the mixing of pointed flows, formation of flotocomplexes (active silt)-(air bubbles) and then their surface-emerging into foam layer, forming in foam Gutter *5*, proceeds.

Then foam product via Horse *11* and further via Sleeve *16* comes inside screw Thickener *13* wherein also sediment simultaneously is supplied in via Sleeve *17*. The mixture of foam product and sediment is dehydrated on account of rotating under Electro-engine *12* effect Screw *15* and then goes into Chamber *21*, being thickened till dehydrated product consistency by the way of additional bend of liquid at the expense of false Bottom *18* with live section from 20 to 60 % and with perforations in the form of squares with side from 0.1 to 0.9 mm. Meanwhile, being separated from sediment, liquid in the form of fugate is being withdrawn from screw thickener via Sleeve *14*. Sediment, dehydrated approximately till residual humidity of 80–90 %, is being withdrawn from Chamber *21* via Sleeve *20*, and liquid — via Sleeve *19*. The choice of screw length, which constitutes in the optimal variant from 25:1 to 35:1 relatively screw diameter, promotes the obtaining of the lowest residual humidity. Thus, as experiments have shown, the optimal value of false bottom constitutes 20–60 % in the limits of common value of Chamber *21* bottom square, and, having quadrate form, perforations of false bottom are optimal with the quadrates' side from 0.1 to 0.9 mm.

Defecated water goes through windows of Wall *6* and Wall *8* with window, closed with net, that does not let into those flotocomplexes which are late to emerge on surface into foam layer. It allows to reach high efficiency of separation of active silt floccules from water at net cell size value, closing the window of Wall *8* in the limits from 0.01 to 0.7 mm. While the use of net which is out of pointed value, the purification efficiency falls down, namely, at cell sizes less than 0.01, water movement resistance rises drastically, and, at sizes more than 0.7, flotocomplexes particles-bubble slip away. Whereat, additionally, ferromagnetic particles can linger on the net because the net can be magnetic as a material.

Further, defecated water is drawn via Device *9* of liquid level regulation and Sleeve *10*, out of Shell *1* of flotoharvester, and can be used by purpose.

Offered flotoharvester usage allows to raise the efficiency of separation of active silt floccules from water approximately on 15–20% and increase sediment thickening degree by 1.5–2 times.

Complexes interaction between each other and their transfer from one state to another constitutes the basis for models of processes, proceeding in flotoharvester.

Product of one stage passes to next stage in consequent processes.

Consequent processes can contain from two stages to several thousand ones.

For instance, simple case of model of multistage flotation.

$$A \xrightarrow{k_1} B \xrightarrow{k_2} C.$$

For initial conditions $t = 0$, $CA = a_0$, $CB = CC = 0$, kinetic equations will be the following:

$$\begin{cases} \dfrac{dC_A}{dt} = -k_1 C_A; \\ \dfrac{dC_B}{dt} = k_1 C_A - k_2 C_B. \end{cases}$$

Solution:

$$C_A = a_0 e^{-k_1 t};$$

$$C_B = \frac{C_A k_1}{k_2 - k_1} \left(e^{-k_1 t} - e^{-k_2 t} \right).$$

At

$$t = t_m = \frac{1}{k_2 - k_1} \ln \frac{k_2}{k_1},$$

intermediate product concentration reaches the maximum:

$$C_{B_{max}} = a_0 \left(\frac{k_2}{k_1} \right)^{\frac{k_2}{k_1 - k_2}}.$$

End product concentration:

$$C_C = a_0 \left(1 - \frac{k_2}{k_2 - k_1} e^{-k_1 t} + \frac{k_1}{k_2 - k_1} e^{-k_2 t} \right).$$

Let consider the case of model for mega-aerofloccules formation in the process of active silt flotation:

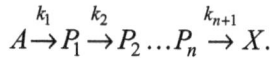

$$A \xrightarrow{k_1} P_1 \xrightarrow{k_2} P_2 \ldots P_n \xrightarrow{k_{n+1}} X.$$

For initial conditions:

$$t = 0,$$

$$C_A = a_0,$$

$$C_{P_1} = C_{P_2} = \ldots C_{P_n} = 0,$$

the solution for intermediate products will be the following:

$$C_{P_i} = a_0 \prod_{j=1}^{i} k_j \sum_{j=1}^{i+1} \frac{e^{-k_j t}}{\prod_{m=1}^{i+1}(k_m - k_j)},$$

where $m = j$.

There represents an interest the case of wastewater conditioning with the use of reagent — collector B, interacting with initial pollution A. Whereat, P — state of flotocomplex "particle–air bubble", and D — state in foam layer.

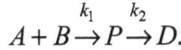

$$A + B \xrightarrow{k_1} P \xrightarrow{k_2} D.$$

For initial conditions $t = 0$, $CA = a_0$, $CB = b_0$, $CP = CD = c_0$, kinetic equations will be the following:

$$\begin{cases} \dfrac{dC_A}{dt} = \dfrac{dC_B}{dt} - k_1 C_A C_B; \\[2mm] \dfrac{dC_P}{dt} = k_1 C_A C_B - k C_P. \end{cases}$$

Solution:

$$C_A = a_0 \frac{b_0 - a_0}{b_0 e^{-k_1(b_0 - a_0)t} - a_0};$$

$$C_B = b_0 - (a_0 - C_A);$$

$$C_P = -\frac{a_0}{1 + a_0 k_1 t} + b_0 e^{-k_2 t} + \frac{k_2}{k_1} e^{\left(-k_1 k_2 a_0^{-1} + k_2 t\right)} \cdot \left(l_i \cdot e^{\left(k_1^{-1} k_2 a_0^{-1} + k_2 t\right)} - l_i \cdot e^{k_1^{-1} k_2 a_0^{-1}} \right).$$

where $l_i x = \int_0^x \dfrac{dz}{\ln z}$ — integral logarithm.

There are also interesting the cases of microflotocomplexes formation and their coalescence by the scheme:

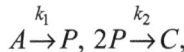

$$A \xrightarrow{k_1} P, \; 2P \xrightarrow{k_2} C,$$

where A — original pollution;

P — non-floating microflotocomplex (flotocomplex which is late to emerge on surface into foam layer while being in flotation zone because of small lifting force of microbubble);

$2P$ — floating microflotocomplex (flotocomplex which can surface-emerge into foam layer in flotation zone during prolonged time of being in the zone).

For initial conditions $t = 0$, $CA = a_0$, $CP = CC = 0$, solution will be the following:

$$C_P = a_0 \left(\frac{e^{-xt}}{x}\right)^{\frac{1}{2}} \cdot \frac{iJ_1\left[2i\left(xe^{-xt}\right)^{\frac{1}{2}}\right]}{J_0\left[2i\left(xe^{-xt}\right)^{\frac{1}{2}}\right]} + \frac{\beta H_1^{(1)}\left[2i\left(xe^{-xt}\right)^{\frac{1}{2}}\right]}{\beta H_0^{(1)}\left[2i\left(xe^{-xt}\right)^{\frac{1}{2}}\right]},$$

where $x = a_0 \dfrac{k_2}{k_1}$;

$$\beta = \frac{iJ_1\left(2i\sqrt{x}\right)}{H_1^{(1)}\left(2i\sqrt{x}\right)};$$

i — imaginary unit;

J_0, J_1 — Bessel function;

$H_0^{(1)}, H_1^{(1)}$ — Hankel function.

Let consider floating of pollutions in the form of microflotocomplexes according to the scheme:

$$2A \xrightarrow{k_1} P \xrightarrow{k_2} C,$$

where A — initial pollution in the form of non-floating microflotocomplex;
  P — floating microflotocomplex;
  C — state in foam layer.

For initial conditions $t = 0$, $CA = a_0$, $CP = CC = 0$, solution will be the following:

$$C_P = \frac{1}{2} a_0 \left\{ e^{-\eta(\tau-1)} + \eta e^{-\eta \tau} E_i\left(\eta \tau\right) - E_i\left(\eta\right)\tau^{-1} \right\},$$

where $\eta = \dfrac{k_2 a_0}{k_1}$;

$\tau = 1 + a_0 k_1 t$;

$C_{P_{max}} = \dfrac{1}{2} \eta \tau_{max}^2 a_0$;

E_i — integral exponential function.

60

Let consider floating of sub-microflotocomplexes under the scheme:

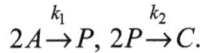

$$2A \xrightarrow{k_1} P, \; 2P \xrightarrow{k_2} C.$$

where A — initial pollution in the form of non-floating sub-microflotocom-
plex (flotocomplex which can't emerge on surface into foam layer during whatever long time of being in flotation zone because of small lifting force of microbubble);

P — non-floating microflotocomplex;

C — state in foam layer.

For initial conditions $t = 0$, $CA = a_0$, $CP = CC = 0$, solution will be the
following:

$$C_P = \frac{a_0}{2\eta\tau}\left[(\alpha+1) - 2\alpha\left(1 + \frac{\alpha-1}{\alpha+1}\tau\alpha\right)^{-1}\right],$$

$$C_{P_{max}} = a_0 \tau_{max}\left(\frac{1}{2\eta}\right)^{\frac{1}{2}},$$

where $\eta = \dfrac{k_1}{k_2}$,

$\tau = 1 + a_0 k_1 t$,

$\alpha = (1 + 2\eta)^{\frac{1}{2}}$.

There is a big practical interest in the consideration of the processes with
limiting stage.

If the process includes a row of consecutive stages and speed constant
of one of the stages is much less than speed constants of other stages, such
stage will be the limiting one, namely, it determines the speed of all of the
process. Just said is characteristic, for instance, for purification processes
with the use of ionic flotation, proceeding by the scheme:

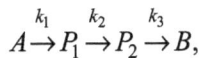

$$A \xrightarrow{k_1} P_1 \xrightarrow{k_2} P_2 \xrightarrow{k_3} B,$$

where A — initial pollution (metal ion);

B — complex "metal ion — collector";

C — flotocomplex "metal ion — collector — gas bubble";

X — state in foam layer.

If $k_1 < k_2$ and $k_1 < k_3$, the limiting stage is the first one; and if $k_2 < k_1$ and
$k_2 < k_3$, the limiting stage is the second one and so on. When orders of consecutive stages are different, there should be compared not speed constants but

speeds of consecutive stages. For instance, in the case of process, proceeding by the scheme:

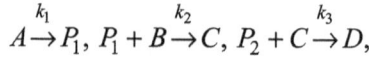

$$A \xrightarrow{k_1} P_1, \; P_1 + B \xrightarrow{k_2} C, \; P_2 + C \xrightarrow{k_3} D,$$

the second stage will limit under conditions k2CB<k1 and k2CB<k3CC.

Scheme of sequential-parallel processes of flotation and sedimentation is:

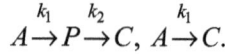

$$A \xrightarrow{k_1} P \xrightarrow{k_2} C, \; A \xrightarrow{k_1} C.$$

For initial conditions $t = 0$, $CA = a_0$, $CP + CC = 0$, kinetic equations are:

$$\begin{cases} \dfrac{dC_A}{dt} = -\left(k_1 + k_1'\right)C_A; \\[2mm] \dfrac{dC_P}{dt} = k_1 C_A - k_2 C_P; \\[2mm] \dfrac{dC_C}{dt} = k_1' C_A + k_2 C_P. \end{cases}$$

Solutions:

$$C_P = \frac{a_0 k_1}{k_2 - k_1 - k_1'}\left[e^{-(k_1+k_1')t} - e^{-k_2 t}\right],$$

$$C_P = \frac{a_0 k_1'}{k_1 + k_1'}\left(1 - e^{-(k_1+k_1')t}\right) + \frac{a_0 k_1}{k_2 - k_1 - k_1'}\left\{1 - e^{-k_2 t} + \frac{k_2}{k_1 + k_1'}\left(1 - e^{-(k_1+k_1')t}\right)\right\},$$

$$\frac{dC_C}{d\Delta C_A} = \frac{k_1}{k_1 + k_1'} + \frac{k_2}{k_1 + k_1'}\frac{C_P}{C_A}.$$

Let consider the process of floccule formation with the use of coagulants and flocculants which proceeds by the scheme:

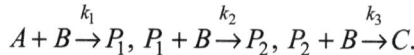

$$A + B \xrightarrow{k_1} P_1, \; P_1 + B \xrightarrow{k_2} P_2, \; P_2 + B \xrightarrow{k_3} C.$$

Initial pieces of kinetic curves are described by equations:

$$\begin{cases} \dfrac{dC_{P_1}}{dt} = k_1 C_A C_B, \; C_{P_1} \cong k_1 a_0 b_0 t; \\[2mm] \dfrac{dC_{P_2}}{dt} = k_2 C_A C_{P_1}, \; C_{P_2} \cong \dfrac{1}{2}k_1 k_2 a_0 b_0^2 t^2; \\[2mm] \dfrac{dC_C}{dt} = k_3 C_B C_{P_2}, \; C_{P_2} \cong \dfrac{1}{6}k_1 k_2 k_3 a_0 b_0^3 t^3, \end{cases}$$

and ratios between concentrations — by equations:

$$-\frac{dC_{P_1}}{dC_A} = \frac{k_1 C_A C_B - k_2 C_B C_{P_1}}{k_1 C_A C_B} = \frac{\eta_1 C_A - C_{P_1}}{\eta_1 C_A},$$

$$-\frac{dC_{P_2}}{dC_A} = \frac{k_2 C_B C_{P_1} - k_3 C_B C_{P_2}}{k_1 C_A C_B} = \frac{\eta_1 C_{P_1} - C_{P_2}}{\eta_1 \eta_2 C_A}.$$

Solutions:

$$C_{P_1} = \frac{\eta_1}{1-\eta_1} C_A + \frac{\eta_1}{\eta_1-1}\left(\frac{C_A}{a_0}\right)^{\frac{1}{\eta_1}} a_0;$$

$$\frac{C_{P_2}}{a_0} = \frac{\eta_1 \eta_2}{(1-\eta_2)(\eta_1-1)}\left(\frac{C_A}{a_0}\right)^{\frac{1}{\eta_1}} + \frac{\eta_1 \eta_2 C_A}{a_0(1-\eta_1)(1-\eta_1\eta_2)} +$$

$$+ \frac{\eta_1 \eta_2^2}{(1-\eta_1)(1-\eta_1\eta_2)}\left(\frac{C_A}{a_0}\right)^{\frac{1}{\eta_1\eta_2}},$$

where $\eta_1 = \dfrac{k_1}{k_2}, \eta_2 = \dfrac{k_2}{k_3}$.

In the series of cases, the floating of microflotocomplexes can be described by the scheme:

$$A \xrightarrow{k_1} P, \; A + P \xrightarrow{k_2} C.$$

For initial conditions $t = 0$, $CA = a_0$, $CP = CC = 0$, kinetic equations will be the following:

$$\begin{cases} \dfrac{dC_A}{dt} = -(k_1 C_A + k_2 C_A C_P); \\ \dfrac{dC_P}{dt} = k_1 C_A - k_2 C_A C_P). \end{cases}$$

Solutions:

$$C_P + 2\eta \ln \frac{\eta - C_P}{\eta} = C_A - a_0;$$

$$\frac{C_P}{\eta} + 2\ln\left(1 - \frac{C_P}{\eta}\right) = -\frac{a_0}{\eta}\left(1 - \frac{C_A}{a_0}\right).$$

Stationary value of intermediate product concentration:

$$C_{P_{stat}} = \eta = \frac{k_1}{k_2}.$$

In the case of microflotocomplexes flotation with reagents (coagulants), we have:

$$A + B \xrightarrow{k_1} P, \; P + B \xrightarrow{k_2} C.$$

For initial conditions $t = 0$, $CA = a_0$, $CB = b_0$, $CP + CC = 0$, kinetic equations look as following:

$$\begin{cases} \dfrac{dC_A}{dt} = -k_1 C_A C_B; \\[2mm] \dfrac{dC_B}{dt} = -(k_1 C_A C_B + k_2 C_B C_P); \\[2mm] \dfrac{dC_P}{dt} = k_1 C_A C_B - k_2 C_B C_P). \end{cases}$$

Solution:

$$\frac{C_P}{C_A} = \frac{1}{\eta - 1}\left[1 - \left(\frac{C_A}{a_0}\right)^{n-1}\right],$$

where $\eta = \dfrac{k_2}{k_1}$.

Maximal concentration of intermediate product:

$$C_{P_{max}} = a_0 \eta^{\varepsilon},$$

where $\varepsilon = \dfrac{\eta}{1 - \eta}$ for $h < 1$,

$C_{P_{max}} = \dfrac{a_0}{e}$ for $h = 1$.

Let consider process of coalescence and floating of microflotocomplexes, the process is described by the scheme:

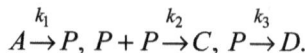

$$A \xrightarrow{k_1} P, \; P + P \xrightarrow{k_2} C, \; P \xrightarrow{k_3} D.$$

For initial conditions $t = 0$, $CA = a_0$, $CP = CC = CD = 0$, kinetic equations will be the following:

$$\begin{cases} \dfrac{dC_A}{dt} = -k_1 C_A; \\[3mm] \dfrac{dC_P}{dt} = k_1 C_A - k_2 C_P^2 - k_2 C_P). \end{cases}$$

Solutions:

if $\dfrac{k_3}{k_1}$ — an integer,

$$C_P = \frac{k_1 \sqrt{\eta_2}}{k_2} \cdot \frac{N_{-\eta_1}\left(2\sqrt{\eta_2}\right) J_{-\eta_1}\left(2\sqrt{\eta_2}\,\sigma\right) - J_{-\eta_1}\left(2\sqrt{\eta_2}\right) N_{-\eta_1}\left(2\sqrt{\eta_2}\,\sigma\right)}{N_{-\eta_1}\left(2\sqrt{\eta_2}\right) J_{1-\eta_1}\left(2\sqrt{\eta_2}\,\sigma\right) - J_{-\eta_1}\left(2\sqrt{\eta_2}\right) N_{1-\eta_1}\left(2\sqrt{\eta_2}\,\sigma\right)};$$

$$\eta_1 = 1 - \frac{k_3}{k_1}, \quad \eta_2 = a_0 \frac{k_2}{k_1}, \quad \sigma = e^{-\frac{1}{2}k_1 t},$$

if $\dfrac{k_3}{k_1}$ — not an integer,

$$C_P = \frac{k_1 \sqrt{\eta_2}}{k_2} \cdot \frac{J_{\eta_1}\left(2\sqrt{\eta_2}\right) J_{-\eta_1}\left(2\sqrt{\eta_2}\,\sigma\right) - J_{-\eta_1}\left(2\sqrt{\eta_2}\right) J_{\eta_1}\left(2\sqrt{\eta_2}\,\sigma\right)}{J_{\eta_1}\left(2\sqrt{\eta_2}\right) J_{1-\eta_1}\left(2\sqrt{\eta_2}\,\sigma\right) - J_{-\eta_1}\left(2\sqrt{\eta_2}\right) J_{1-\eta_1}\left(2\sqrt{\eta_2}\,\sigma\right)},$$

where J — Bessel function;
N — Neuman function.

Let consider consecutive processes with equilibrium stages.
For instance, floating with reversible first stage is described by the scheme:

$$A \overset{k_1,\ k_2}{\longleftrightarrow} B \overset{k_3}{\to} C,$$

for initial conditions $t = 0$, $CA = a_0$, $CB = CC = 0$, kinetic equations look in the following way:

$$\begin{cases} \dfrac{dC_A}{dt} = -k_1 C_A + k_2 C_B; \\[3mm] \dfrac{dC_B}{dt} = k_1 C_A - (k_2 + k_3) C_B; \\[3mm] \dfrac{dC_C}{dt} = k_3 C_B. \end{cases}$$

Solutions:

$$C_A = a_0 \left\{ \frac{\lambda_2 - k_1^{-1} k_3}{\lambda_2 (\lambda_2 - \lambda_3)} e^{-\lambda_2 k_1 t} + \frac{k_1^{-1} k_3 - \lambda_3}{\lambda_3 (\lambda_2 - \lambda_3)} e^{-\lambda_3 k_1 t} \right\},$$

$$C_B = \frac{a_0}{\lambda_2 - \lambda_3} \left\{ e^{-\lambda_3 k_1 t} - e^{\lambda_2 k_1 t} \right\},$$

where

$$\lambda_2 = \frac{1}{2}(\alpha + \beta), \ \alpha = 1 + \frac{k_2}{k_1} + \frac{k_3}{k_1}, \ \lambda_3 = \frac{1}{2}(\alpha - \beta), \ \beta = \left(\alpha^2 - \frac{4k_3}{k_1} \right)^{\frac{1}{2}}.$$

Floating with reversible second stage is described by the scheme:

$$A \xrightarrow{k_1} B \overset{k_2, \ k_3}{\leftrightarrow} C.$$

For initial conditions $t = 0$, $CA = a_0$, $CB = CC = 0$, $t \to \infty$, $CB \to CB$, $CC \to CC$, kinetic equations look in the way:

$$\begin{cases} \dfrac{dC_A}{dt} = -k_1 C_A; \\ \dfrac{dC_B}{dt} = k_1 C_A + k_3 C_C - k_2 C_B; \\ C_A = a_0 e^{-k_1 t}. \end{cases}$$

Solutions:

$$C_B = \frac{C_{B\infty}}{k_1 + k_2 - k_3} \left\{ a_0 (k_1 - k_2) e^{-k_1 t} - \left[k_2 C_{B\infty} - C_{C\infty} (k_3 - k_1) \right] e^{-(k_2 + k_3) t} \right\};$$

$$C_C = \frac{C_{C\infty}}{-(k_1 + k_2 - k_3)} \left\{ a_0 k_2 e^{-k_1 t} - \left[k_2 C_{B\infty} - C_{C\infty} (k_3 - k_1) \right] e^{-(k_2 + k_3) t} \right\}.$$

Floating with collector and flocculant proceeds by the scheme:

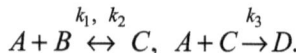

$$A + B \overset{k_1, \ k_2}{\leftrightarrow} C, \ A + C \xrightarrow{k_3} D.$$

For initial conditions $t = 0$, $CA = a_0$, $CB = b_0$, $CC = CD = 0$, kinetic equations will be the following:

$$
\begin{cases}
\dfrac{dC_A}{dt} = -k_1 C_A C_B + k_2 C_C - k_3 C_A C_C; \\[2mm]
\dfrac{dC_B}{dt} = -k_1 C_A C_B + k_2 C_C; \\[2mm]
\dfrac{dC_P}{dt} = k_1 C_A C_B - k_2 C_C - k_3 C_A C_C.
\end{cases}
$$

If at $t = 0$ $CC = c_0$ and concentrations of the rest of products are equal to zero and whereat equilibrium concentration is quickly establishing

$$
C_A = \frac{C_C k_2}{C_B k_1},
$$

then

$$
\frac{dC_C}{dt} = \frac{2C_C^{\,2} k_2 k_3}{C_B k_1} \quad \text{and} \quad \frac{dC_C}{W_C} = \frac{k_1}{2k_2 k_3} \cdot \frac{C_B}{C_C}.
$$

Floating with the use of various reagents is described by the scheme:

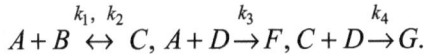

$$
A + B \overset{k_1,\ k_2}{\leftrightarrow} C,\ A + D \overset{k_3}{\to} F,\ C + D \overset{k_4}{\to} G.
$$

Let's suppose that stationary concentration C in the system is being quickly reached, then

$$
C_C = \frac{k_1 C_A C_B}{k_2 + k_4 C_A};
$$

$$
W = \frac{dC_A}{dt} = \frac{k_1 k_4 C_A C_B C_D}{k_2 + k_4 C_A} + k_3 C_A C_D;
$$

$$
k_{exp} = \frac{W}{C_A} = \frac{\Delta \ln C_A}{\Delta t} = \left(\frac{k_1 k_4 C_B}{k_2 + k_4 C_D} + k_3 \right) C_D = k' C_D;
$$

$$
\lim_{C_A \to 0} k' = k_3,\ \lim_{C_D \to 0} k' = \frac{k_1 k_4 C_B}{k_2} + k_3,\ \lim_{C_D \to \infty} \overline{k_{exp}} = k_1 C_B + k_3 C_D.
$$

Offered solutions can be used in the practice of flotation thickening of active silt, in particular, at more precise calculation of flotation machines and apparatus with various aeration systems. Whereat, considered cases embrace the majority of processes of wastewater flotation purification that is especially important in the use of combined flotation facilities, including flotoharvesters [49–52] which are also applied in the practice of wastewater purification.

In the simplest variant of flotoharvester, it can be accomplished in the form of flotodecanter a fuller separation of active silt from water in comparison with the use of regular technique, in particular, to increase the degree of extraction (separation) of active silt on 10–15% and to promote a specific hydraulic load on 20–25% in comparison with corresponding indicators of known flotodecanters.

Trial and introduction of similar flotodecanter have been accomplished in the practice of wastewater purification (Fig. 4.5).

Fig. 4.5. Photo of industrial sample of flotodecanter

Testing and introduction of flotodecanter of pointed construction have shown stability of its work and prospects of its usage in wastewater purification practice.

High concentration of active silt changes its properties — it lessens a specific speed of organic compounds oxidation, a capability of silt mixture for division and silt sedimentation. Nevertheless, at the exclusion of these drawbacks, the method of biological purification intensification of active silt excessive concentration doesn't allow, in comparison with usual approach, to increase the depth of purification according to COD (Chemical Oxygen Demand), to lessen the content of specific pollutions in the purified water which usually hardly undergo biological purification.

Different variants of flotation process intensification with the use of multistage model are considered by us in some works and describe flotation process pursued in particular conditions [36–42].

4.3 FLOTATION PROCESS GENERALIZED MODELS

Together with aforementioned flotation and accompanying processes, generalization of description of such difficult processes represents an interest. Whereat, we should note that generalization relates to description of processes which multiple particles and bubbles participate in.

Description generalization of complicated processes consideration represents big practical interest [15, 50] while which fulfillment, there proceeds the formations of aerofloccules; aerofloccules include hydrophobic as well as hydrophilic particles.

Let's consider the usage of generalized model applied to wastewater flotation purification with the preliminary wastewater conditioning with reagents. While the conditioning, in comparison with regular flotation, where the formation of flotocomplex (system "hydrophobic pollution particle — air bubble") proceeds, more complicated complex forms — it is an aerofloccule which center are small air bubbles; hydrophobic as well as hydrophilic particles are attached to the air bubbles. It is necessary the fulfillment of 2 conditions simultaneously: effective mixing of reagents with being purified water and air suction and dispersal. The necessity to obtain smaller air bubbles is noted in many domestic and foreign works.

In general case, during wastewater conditioning accomplishment, reversal processes can proceed, namely, aerofloccules formation, flotocomplex falling out from foam layer, and also solid phase particles falling out from foam layer into aeration zone. Offered model of conditioning of wastewaters, containing hydrophobic and hydrophilic pollutions, in view of reversibility of processes is shown on Fig. 4.6.

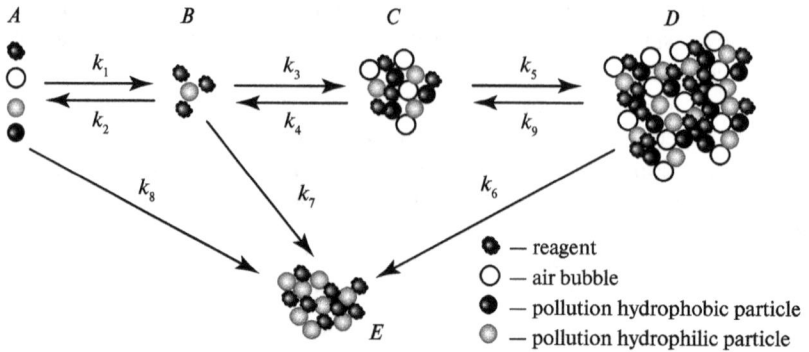

Fig. 4.6. Model of flotation process of wastewater purification with wastewater conditioning

State A on the shown figure — original state of pollution particles, reagent is in the form of metals hydroxide and air bubbles, state B — formation of complex (pollution particle)-reagent, state C — aerofloccule formation, state D — particle state in foam layer, state E — particle state in sediment. Constants k_1, k_2, k_3, k_4, k_5, k_6, k_7, k_8 characterize transfers from one state to another.

Model, presented on Fig. 4.6, illustrates processes proceeding in the conditioning chamber. On the primary stage (state A), being purified water with hydrophilic and hydrophobic pollutions, is in the conditioning chamber, reagent supply and air suctions are not being fulfilled. Whereat, the largest pollutions are starting to fall out into sediment (state E). Then there are supplied the reagents (coagulant and flocculant), which molecules during mixing are starting to interact with the pollution particles (state B), the complexes "(pollution particle) — reagent" are formed. Whereat, also sedimentation of the heaviest complexes is going on. When mixer, which mixes the reagents and wastewater, reaches big number of turns, air starts to be sucked. Occurring in rotation zone of mixer, the air is crashed into multiple tiny bubbles. Together, active mixing of wastewater proceeds already not just with the reagents but with fine-dispersed air bubbles which not only pollutions hydrophobic particles but also treated with reagents hydrophilic pollutions start to stick to. Thus, large complex forms, called aerofloccule (state C), forms. Then aerofloccules on account of Archimedian force, go upwards where they are gathered in the form of foam layer (state D). In the foam layer, with the time, the air bubbles can be disrupted that leads to that the part of caught pollutions fells out into sediment (state E).

70

Given model is described by the following differential equation system:

$$\begin{cases} \dfrac{dC_A}{dt} = -k_1 C_A + k_2 C_B - k_8 C_A; \\[2mm] \dfrac{dC_B}{dt} = k_1 C_A - k_2 C_B - k_3 C_B + k_4 C_C - k_7 C_B; \\[2mm] \dfrac{dC_C}{dt} = k_1 C_A C_B - k_2 C_C - k_3 C_A C_C; \\[2mm] \dfrac{dC_D}{dt} = k_5 C_B - k_6 C_C; \\[2mm] \dfrac{dC_E}{dt} = k_8 C_A + k_7 C_B + k_6 C_D. \end{cases} \tag{4.1}$$

Constant k_1 characterizes "particle-reagent" complex formation.

Constant k_2 characterizes "particle-reagent" complex crash.

Constant k_3, characterizing aerofloccule formation, is defined from the relation:

$$k_3 = \frac{1.5qE}{k_0 \bar{D}}, \tag{4.2}$$

where q — barbotage speed, m^3/(m$^2\cdot$s);

E — efficiency of particle's capture by surface-emerging air bubble during flotation (dimensionless);

$\bar{D}$ — average diameter of bubbles in flotation cell, m;

k_0 — polydispersity factor of bubbles (dimensionless).

Particle's capture efficiency by surface-emerging gas bubble during flotation is determined by formula (4.3):

$$E = 0.5 \frac{r_p^{1.6}}{r_b^2} A^{1/6}, \tag{4.3}$$

where r_p — particle radius, m;

r_b — bubble radius, m;

A — Hamaker's constant, J.

Constant k_4 characterizes aerofloccule crash.

Constant k_5 characterizes aerofloccule moving from liquid into foam layer and is defined as:

$$k_5 = \frac{v_{em}}{h}, \qquad (4.4)$$

where v_{em} — flotocomplex lift speed, m/s;

h — distance from aeration zone to foam layer (flotochamber depth), m.

Surface-emerging speed of bubbles can be defined approximately upon formula (5):

$$v_{em} = \frac{\bar{D}^2 g (\rho_w - \rho_a)}{18\mu}, \qquad (4.5)$$

where $g = 9{,}81$ m/s^2 — acceleration of gravity;

ρ_w — water density, g/m^3;

ρ_a — air density, g/m^3;

μ — water dynamic viscosity coefficient, Pa·s.

Constant k_6 characterizes solid phase particles' falling out from foam layer into aeration zone.

Constant k_7 characterizes sedimentation of solid phase particles being in original state.

Constant k_6 characterizes solid phase particles' falling out from foam layer.

Constant k_3, k_5 values were calculated for bubble various diameters in interval of 40–200 microns. The values are presented in Table 4.1.

Table 4.1. Constant k_3, k_5 values depending on bubble diameter

$D \cdot 10^{-6}$, m	k_3, s^{-1}	k_5, s^{-1}
40	0.0574	0.0009
50	0.0294	0.0014
60	0.0170	0.0019
70	0.0107	0.0026
80	0.0072	0.0034
90	0.0050	0.0044
100	0.0036	0.0054
110	0.0027	0.0065
120	0.0021	0.0078
130	0.0017	0.0092
140	0.0013	0.0106
150	0.0011	0.0122
200	0.0004	0.0217

For practical purposes, it is comfortable to apply the chart of numerical solution of differential equation system at initial conditions $t = 0$, $C_A = C_0$, $C_B = C_C = C_D = 0$.

Set parameters of the process are:

$C_0 = 0.001$ g/l; $E = 0,038$; $k_0 = 1$; $D = (40-200) \cdot 10^{-6}$ m;
$q = 4 \cdot 10^{-5}$ m³/(m²·s); $v_{em} = 0.9 \cdot 10^{-3}$ m/s; $h_{em} = 1$ m.

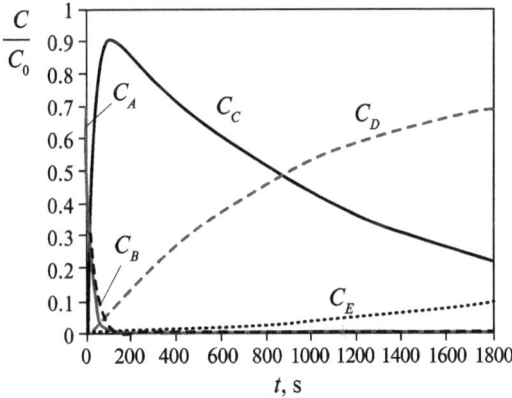

Fig. 4.7. Solution of differential equation system in graphical view (bubble size is $40 \cdot 10^{-6}$ m)

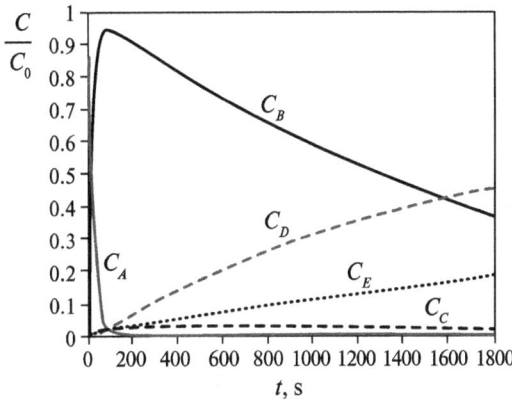

Fig. 4.8. Solution of differential equation system in graphical view (bubble size is $200 \cdot 10^{-6}$ m)

Theoretically calculated constants of the process are:

$k_1 = 0.05$ s⁻¹; $k_2 = 0.0001$ s⁻¹; $k_3 =$ look Table 4.1;
$k_4 = 0.0001$ s⁻¹;
$k_5 =$ look Table 4.1;
$k_6 = 0.0011$ s⁻¹;
$k_7 = 0.0011$ s⁻¹;
$k_8 = 0.0011$ s⁻¹;
$k_9 = 10^{-5}$ s⁻¹.

After analysis of different variants [92–94], numerical method of system solution in Scilab was chosen; Scilab is the package of applied mathematical programs with the possibility to use for technical and scientific calculations [95].

The system solution was obtained in graphical view and presented on Fig. 4.7–4.8.

Presented data on Fig. 4.7–4.8 show how flotation process kinetics changes while bubble size change. Whereat, variations of constants, depending on size of the bubbles, show their significance for flow of flotation process individual stages.

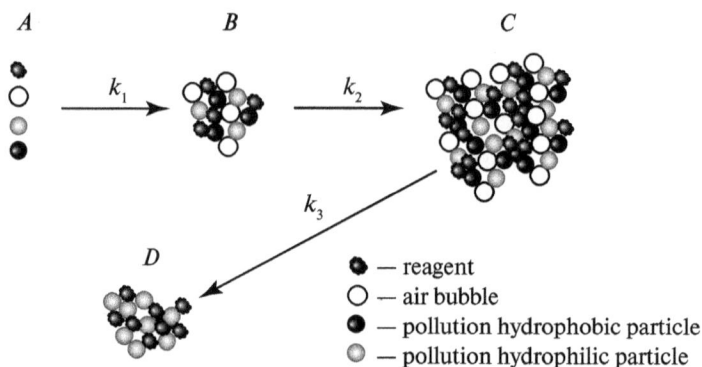

Fig. 4.9. Model of wastewater purification flotation process
with wastewater conditioning; reversibility
is not taken into account

On the beginning stage of purification process, intensive mixing of wastewater with reagents proceeds that leads to formation of "(pollution particle) — reagent" complex (curve C_B) and reduction of concentration of pollutions in initial effluent (curve C_A). Then, via casing holes, an air suction and its further dispersal into tiny bubbles in mixer rotation zone proceeds that leads to aerofloccule concentration increase (curve C_C). It should be noted that the smaller air bubble the more intensively aerofloccule formation proceeds. Then they surface-emerge into foam layer and aerofloccule concentration in effluent lessens. Pollution concentration in foam layer (curve C_D) and in sediment (curve C_E) increases during all of the process and, besides, their sum strives to horizontal asymptote with function value equal to pollution initial concentration.

According to the shown model, there can be found the purification time, for instance, for air bubble of $40 \cdot 10^{-6}$ m size, to obtain 90% efficiency the purification necessary time constitutes about 600 s.

The reversibility can be ignored in the case of effluent purification from well-floating substances (for instance, mineral oils, oils and some substances with the use of reagents). Meanwhile, as a rule, the stage of formation of "(pollution particle)-reagent" complex goes on very quickly and can be ignored. Hence, wastewater conditioning model may look as it is shown on Fig. 4.9

The process, shown on Fig.4.9, is described by differential equation system (4.6).

74

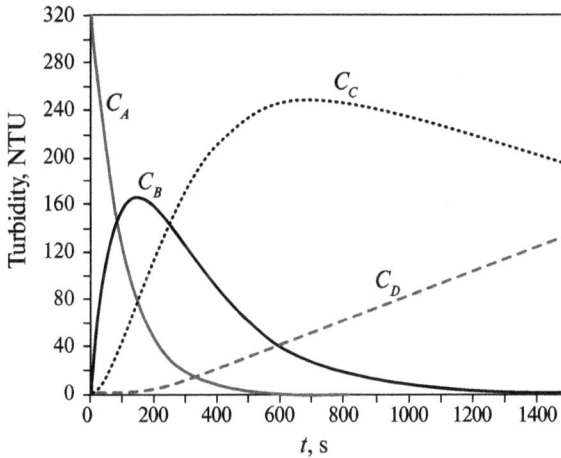

Fig. 4.10. Solution of differential equation system of flotation simplified model in graphical view

In differential equation system (4.6):

$$\begin{cases} \dfrac{dC_A}{dt} = -k_1 C_A; \\[2mm] \dfrac{dC_B}{dt} = k_1 C_A - k_2 C_B; \\[2mm] \dfrac{dC_C}{dt} = k_2 C_B - k_3 C_D; \\[2mm] \dfrac{dC_D}{dt} = k_3 C_D. \end{cases} \qquad (4.6)$$

C_A — initial concentration of pollution particles, reagent and air bubbles, C_B — concentration of aerofloccules, C_C —concentration of particles in foam layer, C_D — concentration of particles in sediment. Constants k_1, k_2, k_3 characterize transfers from one state to another.

Initial conditions for given system of differential equations are the following:

$$t = 0: \ C_A = C_0; \ C_B = C_C = C_D = 0.$$

Transfer constants are calculated according to the following formulas [18]:

$$k_1 = \frac{1.5qE}{k_0\overline{D}};$$ (4.7)

$$k_2 = \frac{\upsilon_{em}}{h};$$ (4.8)

where υ_{em} — speed of aerofloccule surface-emergence;

$$k_3 = \frac{\upsilon_{sed}}{h};$$ (4.9)

where υ_{sed} —speed of aerofloccule sedimentation from foam layer.

For irreversible process case, the graph of change of concentration versus time, shown on Fig. 4.10, was obtained.

Purification time for given efficiency can be defined on the basis of demonstrated graph on Fig. 4.10. For example, efficiency is about 95% at purification time of 10 minutes.

Presented approach is not the only one in the case of consideration of generalized models and represents the simplest variant for the description of hydrophilic and hydrophobic pollutions extraction by flotation way.

5 FLOTATION PROCESSES INTENSIFICATION AND NEW EQUIPMENT

5.1 PRESSURE-HEAD FLOTATION PROCESS INTENSIFICATION WITH THE USE OF CARBON DIOXIDE

To intensify the process by pressure-head flotation is popossible with the help of reagent-treatment. Coagulants, flocculants, detergents are used. It allows to rise aggregate size, to increase their hydrophobicity. It is needed to refer to reagent treatment drawbacks: the increase of foam quantity which entails complexities of dehydration and sediment utilization.

This problem can be settled partially by construction ways, for instance, by facility after the block of fine-layer defecating and filter. It is possible to pursue the secondary saturation of floating media by gas bubbles of large size. The saturation is often held by barbotage. But at such way, besides the coalescence of bubbles, the destruction of flotoaggregates proceeds what with the creation of high speed gradient by big bubble.

We've managed to reach the biggest efficiency in coalescence while pressure-head flotation under conditions when small and big bubbles are formed immediately in the liquid phase. These conditions are possible, for instance, during the use of two or more working liquids with gases of differing solubility.

This idea was firstly told by Ksenofontov B.S. in 1989 and researched in details in the series of his works and especially in monograph [16], and is protected by patent in the wide aspect of its application [25].

According to Ksenofontov B.S. data, the bubble of 0.01−0.05 mm size in average is formed during introduction of one working liquid. Surface-emerging speed of such flotoaggregates constitutes 0.13−0.26 mm/s. Complex aggregate-(gas bubble (hardly soluble gas))-(carbon dioxide gas bubble (easily soluble gas)) is formed at the addition of second liquid, saturated with carbon dioxide. The carbon dioxide bubble coalesces via air bubble.

While such way of flotation, transfer of pollution floating particles into foam proceeds 2−2.5 times quicker. Thickening of foam layer also goes on that is considered in the work in details.

Pressure-head flotation process can be intensified by the way of introduction of the second working liquid [12, 16]. The second working solution introduction − the solution of easily soluble gas brings repeat of the second phase creation — gaseous. For easily soluble gas bubble formation, it is also necessary to spend an energy, whereat, bigger one than during air bubble extraction, because of good gas solubility, so, for second gas extraction, forming centers are needed also. As hardly soluble gas is extract momentarily from solution, so, the air, its bubbles are forming centers of a new gaseous phase — easily soluble gas. Easily soluble gas extraction from solution goes slower than air extraction and, hence, takes more time, but whereat, the enlargement of gas bubble proceeds smoothly, and their stability is not harmed. Flotocomplex is formed as a result, it represents a pollution particle, which volume and surface gas bubbles with sizes approximately 2–3 mm are located in and on.

Flotation extraction of active silt flakes can be described using Ksenofontov's multistage model. Flotation extraction process has three states A, B, C.

State A — active silt flakes being in liquid volume and bubbles are not connected and don't contact. Flake's contact with gas bubble proceeds on the first stage. During sticking, there is being formed flotocomplex flake-bubble (state B), which surface-emerges on account of Archimedean forces. Emerged on the top part of liquid, flotocomplexes form foam layer (state C). Whereat, transfers not only from state A to state B and then to state C but also reversible transfers, correspondingly, from state C to state B and then to state A are also possible. At such approach to the study of flotation extraction and the separation of all of the process on three states, generally, all the process can be described by the following equation system:

$$\begin{cases} \dfrac{dC_A}{dt} = -k_1 C_A + k_2 C_B + k_5 C_C - k_6 C_A; \\[2mm] \dfrac{dC_B}{dt} = k_1 C_A - k_2 C_B - k_3 C_B + k_4 C_C; \\[2mm] \dfrac{dC_c}{dt} = -k_5 C_C + k_6 C_A + k_3 C_B - k_4 C_C, \end{cases} \qquad (5.1)$$

where C_A, C_B, C_C — active silt concentrations, correspondingly in states, A, B, C;
k_1, k_2 — transfer speed constants from state A to state B and back;
k_3, k_4 — transfer speed constants from state B to state C and back;
k_5, k_6 — transfer speed constants from state C to state A and back.

Flotation process kinetic constants k_1–k_6 definition constitutes rather difficult task, nevertheless, namely them allow to pursue the detailed investigations and fulfill scientifically substantiated calculation of apparatuses. The calculation of the constants during flotation have been made with the use of two working solutions.

78

Ksenofontov has offered calculation expressions for all constants except for k_1.

Constant k_1 characterizes the process of particle-bubble flotocomplex formation at the earliest moment of its forming — the process of active silt flake capture by bubble. Mathematical dependence for k_1 constant definition for pressure-head flotation method is absent in the literature, so it can be determined approximately from the formula:

$$k_1 = K \cdot \frac{E}{t}, \qquad (5.2)$$

where K — proportionality coefficient;

E — flotocomplex formation probability;

t — time, s.

As pressure-head flotation method differs by high probability of flotocomplex formation, so

$$E \approx 1. \qquad (5.3)$$

Expression for constant k_1 definition can be presented also in the following form:

$$k_1 = K \cdot \frac{1}{t}. \qquad (5.4)$$

Flotocomplex formation in the case of silt thickening with the use of two working solutions goes in the same way as in the pressure-head flotation traditional method. Air is extracted from liquid at the expense of pressure drastic drop during high-speed outflow of jet of water-air mixture from saturator into flochamber. This air in the form of microbubbles spawns on active silt flakes. Easy-soluble gas solution causes impact on gas bubble growth much later — namely, when complex particle-(air bubble) is already formed. It's been also confirmed in experiments where gas saturated solutions with various solubility have been introduced into flotochamber with differring consistency [19]. If to introduce initially air solution into flotochamber and then easy-soluble gas solution, flotation process begins immediately and is gradually accelerated. The same is observed during simultaneous supply of gases. If to introduce easy-soluble gas firstly into flotochamber and air solution only after that, the flotation process will start only after the introduction of air solution. These experiments confirm that hard-soluble gases are extracted from solutions at the initial moment. Thus, it is seen that at the initial stage, easy-soluble gas doesn't have any impact on the process of particle sticking with bubble. It means that constant k_1 doesn't change as a result of second working solution usage.

Constant k_2 characterizes particle-bubble complex destruction. To analyze destructing processes of the complex, it is necessary to investigate the equation of balance of forces, acting on the bubble:

$$F_1 = F_2 + m_p g + f. \tag{5.5}$$

where F_1 and F_2 are capillary forces of sticking and detachment, correspondingly;

m_p — particle mass;

f — hydrodynamical force of detachment.

Constant k_2 can be determined from the relation [66]:

$$k_2 = n \rho \, v^m G^p \frac{d_\tau M^2}{EN}, \tag{5.6}$$

where G — speed gradient;

E — efficiency of active silt flake capture by surface-emerging bubble of gas;

N — parameter, characterizing stability of binding between flakes and gas bubbles;

M — ratio of flake diameter to bubble diameter;

d — diameter of active silt particle;

v — kinematic viscosity of suspension;

n — concentration of flotocomplexes particle-bubble;

ρ — liquid density;

p, m — coefficients ($1 \le p \le 2; m = 2/3$).

As the efficiency of flotocomplex formation is close to 100%, the expression (4.6) can be presented in the form:

$$k_2 = n \rho \, v^m G^p \frac{d_\tau M^2}{N}. \tag{5.7}$$

During pressure-head flotation, in flotochamber, the situation is calm, regime is laminar because of slow lift of microbubbles, so k_2 aims to zero. Experimental research shows that after introduction of carbon dioxide solution and extraction of gas itself, the bubbles, comprising flotocomplexes, grow and, accordingly, lifting speed increases. N — parameter characterizing the strength of binding between particle and gas bubble, lessens, M — ratio of particle diameter to bubble diameter, increases. In this case, constant k_2 can grow some in comparison with flotation regular regime. Whereat, to define constant new value is rather difficult because of the absence of data, characterizing hydrodynamical situation nearby the growing bubble and moving flotocomplex. Formula 5.7 includes speed gradient value, surely, while media turbulence increase, the speed gradient G, causing the biggest influence on constant k_2 behavior, will grow itself, but to evaluate this increase quantitively is very difficult.

Constant k_3 characterizes the transport of particle-bubble complexes from liquid into foam layer and is defined with the help of the relation

$$k_3 = \frac{\upsilon_{em}}{h}, \qquad (5.8)$$

where υ_{em} — lifting speed of flotocomplex bubble-particle;
  h — distance from aeration zone to foam layer (flotochamber depth).

In the majority of cases, υ_{em} can be defined by the formula:

$$\upsilon_{em} = \frac{g\left[d_g^3\left(\rho_s - \rho\right) - d_g^3\left(\rho - \rho_g\right)\right]}{18\mu d_g}, \qquad (5.9)$$

where g — acceleration of gravity;
  ρ_s, ρ_g — density of solid and gaseous phases;
  μ — liquid dynamical viscosity;
  d_g — gas bubble diameter.

The introduction of easy-soluble gas solution leads to significant raise of bubble sizes, and, hence, as formula 5.9 shows, to lifting speed increase that constitutes the most important indicator of the process intensification. So, it is necessary to study in details the change in bubbles sizes and in speeds of their lift.

The air, dissolved in pressure-head reservoirs (saturators), is extract from the water during pressure decrease, that is used in pressure-head flotation facilities.

Reliable sticking of floating particles with air bubbles is provided at the expense of that, at pressure-head flotation, bubble size is minimal and is extracted directly on being retrieved particle.

Minimal size of air bubble, being formed during depressurization, is proportional to surface tension on a phase boundary and inversely proportional to pressure drop (the larger depressurization of saturation pressure and atmospheric pressure the smaller bubble size).

$$R_{min} = \frac{2\sigma_{g.l}}{P_1 - P_{2\,min}},$$ (5.10)

where R_{min} — bubble minimal radius, m;

$\quad \sigma_{g.l.}$ — surface tension on phase boundary gas-liquid, N/m;

$\quad P_1 - P_2$ — pressure drop, Pa.

It is seen from the formula that the smaller surface tension and the larger pressure drop, the smaller size of being created bubble. Surface tension depends on being purified water properties and availability of detergents therein. Taking that into account, it could be envisaged the introduction of additional reagents, lessening surface tension, however, it is extremely unwillingly. It is more acceptable the obtaining of bubble minimal size by the way of pressure drop increase, i.e. by the way of water saturation with air under heightened pressure and further its drastic decrease till atmospheric one.

Formula 5.10 shows estimated value of diameter of the bubble, extracted from liquid while depressurization from saturation pressure till atmospheric one. Whereat, it is not taken into account any interactions between newly created air bubbles (coalescence, dissolution, growth), hydrodynamical movements in liquid layers are not taken into account, phase transitions are not taken into account. Such simplified formula gives just bubble sizes in the initial moment of fleet process of gas extraction from oversaturated solution during depressurization.

Experiments, held with the use of stereomicroscope, show that when bubble was formed, went through changes as a result of interaction and finally became stable, its sizes constituted 0.01–0.05 mm.

Bubble size definition allows to pursue study of lift speeds of bubbles and flotocomplexes.

The speed of established fall of solid ball of radius R in viscous liquid, proceeding under gravity impact, in the limits of Stocks formula applicability is equal to

$$U = \frac{2R^2 g(\rho' - \rho)}{9\eta},$$

where ρ' — ball density;

$\quad \rho$ — liquid density.

The formula is fair for Reynolds number Re $\ll$ 1.

For liquid spherical drop, being in steady uniform fall in other liquid, Stocks formula will be transformed to the form:

$$U = \frac{2R^2 g(\rho' - \rho)(\eta + \eta')}{3\eta(2\eta + 3\eta')},$$

where ρ' и η' — density and dynamic viscosity of liquid, forming the drop.

It is ensued from the dependencies for spherical drop the dependencies for small bubbles of vapor or gas, lifting in liquid, $\rho' = 0$ and $\eta' = 0$. So, bubble movement speed has the form:

$$U = \frac{1}{3} \frac{R^2 g \rho}{\eta}. \tag{5.11}$$

At very large Re numbers, it can be assumed that viscosity influence is seen only in that part of the liquid which moves in immediate proximity from air-flow layer surface of corps (boundary layer). Liquid speed on corps surface equals zero (sticking condition), the speed on external border of boundary layer depends on speed and cross-sizes of crept flow, corps form and size. The liquid can be considered ideal out of boundary layer. If number Re $\ll$ 1, more precise formula of lift speed definition takes gas density into account.

$$w = \frac{2}{9} \frac{R^2 g (\rho l - \rho g)}{\eta}.$$

If number Re $\gg$ 1, value of lift speed of bubble should be determined by the formula:

$$U = \frac{1}{9} \frac{R^2 g \rho}{\eta}.$$

For bubble small sizes, with taking of circulation into account, the speed can be calculated by the formula:

$$w = \frac{1}{3} \frac{R^2 g (\rho l - \rho g)}{\eta}.$$

It should be noted that maximal size of the bubble, lifting in liquid flow, which Stocks formula is fair for, is calculated by the formula:

$$rg_{max} = \sqrt[3]{\frac{9}{2} \frac{\mu_l^2}{\rho_l g (\rho_l - \rho_g)}}.$$

83

Experimental research, aimed on pressure-head flotation intensification by the way of second working solution (created by easy-soluble gas), showed that average sizes of gas bubbles in flotocomplexes constituted 2 mm. For such bubbles, number Re >> 1, and size itself exceeds aforestablished threshold of acceptability for Stocks formula usage. For big bubbles, lift speeds do not already comply with the laws of spherical particle lift, here bubble shape starts to effect, the shape changes under crept flow, striving to accept the shape of ellipse. Big bubble speeds comply with empirical laws that depends on many parameters. Special graphs, made on the basis of experiments, are demonstrated for them. For the flotocomplex, which bubble in has size ~2 mm, the lift speed constitutes several cm/s (up to 5 cm/s). Nevertheless, turbulent component of the process, leading to the destruction of complexes and the fall of particles, also increases significantly whereat.

According to pursued experimental-industrial trials, made by Ksenofontov B.S, it was stated that the process of flotation with the use of CO_2 solution (as a second working solution) accelerates by 2–2.5 times. It is difficult to calculate theoretically the lift speed of new enlarged bubble and, moreover, of new flotocomplex. On the strength of that all process accelerates by 2–2.5 times, we can suppose that also lift speed in this case is 2–2.5 times more, i.e. it is 3.25–5.2 mm/s, and turbulent component does not change.

In this case, average value of the speed can be defined by the relation:

$$V_{av} = \frac{3.25 + 5.2}{2} = 4.225 \, \frac{\text{mm}}{\text{s}}.$$

Constant k_4 characterizes processes proceeding in foam layer. Second working liquid in the form of additional solution does not have noticeable influence on this constant factually. Whereat, flotocomplex possesses rather big bubble of gas, and transfer of all complex into liquid layer is remote. Most probably, fall of the particle from foam layer into state A would proceed, and this process is characterized by constant k_5.

Constant k_5 characerizes the processes of flakes of active silt falling from foam layer into an aeration zone. It's defined by the formula:

$$k_5 = \frac{\upsilon_{sed}}{h}, \tag{5.12}$$

where υ_{sed} — sedimentation speed of solid phase particles, falling from foam layer. For the majority of the practical tasks, considered in the work, υ_{sed} can be calculated by Stocks formula.

It should be noted that well-soluble gases not only gear up the processes of the transfer of particles into foam layer but also increase stability of formed foams, and, hence, raise up their strength to destruction.

Constant k_6 characterizes the probability of active silt flakes transfer from liquid to foam and is determined from the relation (5.13)

$$k_6 = D \frac{\partial}{\partial x} \left\{ \frac{1}{2\sqrt{\pi Dt}} \left[\exp\left(-\frac{(x-h)^2}{4Dt} \right) - \exp\left(-\frac{(x+h)^2}{4Dt} \right) \right] \right\}, \quad (5.13)$$

where t — time;

x — current distance from foam layer border;

D — diffusion coefficient of particles of solid phase in liquid.

Active silt density is more than water density, and, taking in consideration that situation in flotochamber is calm, the probability of transtition of active silt flakes from suspension into foam layer without flotocomplex formation is minimal.

It should be noted in a whole that introduction of second working solution into flotochamber increases gas bubble sizes, and, therefore, raises up the lift speed of flotocomplexes approximately by 1.5–2 times that leads to the intensification of the process of flotation separation of active silt suspension.

5.2 BASES OF WASTEWATER PURIFICATION PROCESSES IN FLOTATION DECANTER

Usage of flotodecanters allows to reach high technological results in many cases. Regime optimization on the basis of used flotodefecating model provides for it. Incidentally, we were developing different models of flotodefecating on the basis of flotation process multistage model, worked out earlier by Ksenofontov B.S. The scheme of flotodecanter without reagent treatment node is presented on Fig. 5.1.

Let's denote that the sum of concentrations of pollutions, fallen out into sediment (C_D), transferred into foam layer (C_C) and concentrations of pollutions, fixated on bubbles (C_B), for any time t equals the difference of concentrations in original wastewater (C_0) and concentrations of pollutions in wastewater for the same time t (C_A). Or:

$$C_B(t_1) + C_C(t_1) + C_D(t_1) = C_0 - C_A(t_1).$$

Flotation process ending is chosen by developer depending on the set task. In the case of usage of just flotation purification, it is feasible to pursue the

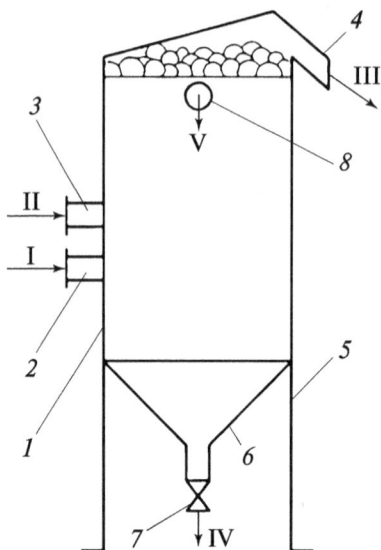

Fig. 5.1. Flotodecanter scheme without reagent treatment node: *1* — flotodecanter shell; *2* — sleeve for working liquid supply; *3* — wastewater supply sleeve; *4* — foam gutter; *5* — racks; *6* — sediment collector; *7* — valve; *8* — sleeve of purified water supply. I — working liquid; II — wastewater; III — foam product; IV — sediment; V — purified water

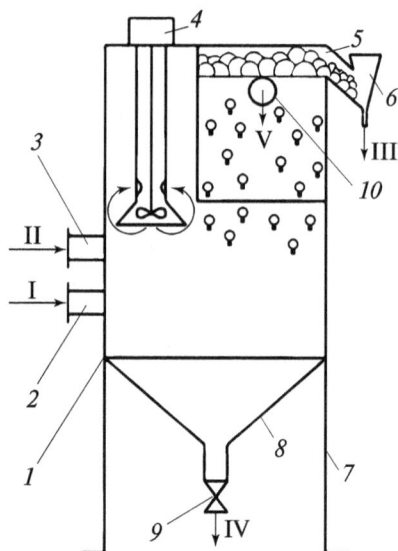

Fig. 5.2. Flotoharvester scheme including reagent treatment node: *1* — flotodecanter shell; *2* — sleeve for working liquid supply; *3* — wastewater supply sleeve; *4* — mixer; *5* — foam gutter; *6* — ejector; *7* — racks; *8* — sediment collector; *9* — valve; *10* — purified water output sleeve. I — working liquid; II — wastewater; III — foam product; IV — sediment; V — purified water

process till such time (t_f) when the change in extraction of pollution particles reaches practically maximal value.

Let's consider other variant of gravitation-flotation process of wastewater purification in flotoharvesters according to the following scheme. Pollution initial particles are submitted to reagent treatment for enlargement and further division from water of hydrophilic pollutions, of sand, in particular, by the way of defecating, and of hydrophobic ones, for instance, mineral oil drops, by flotation (Fig. 5.2).

86

Particular assessment of process various stages can be held with the use of analytical dependencies as well as with the application of analysis calculation methods. Whereat, pursued analysis of models shows that the most applicable approach in given case are calculation methods and only sometimes — analytical ones.

Thus, the submitted approach allows to evaluate all stages of such complex process, comprising reagent treatment of wastewater with further defecating and flotation, and to clarify limiting stages with their consequent intensification and all of the process in a whole.

The sequence of flotation and defecating chambers (first chamber of flotation, second — defecating one or vice versa) is necessary to choose depending on wastewater content and on the basis of experimental research of wastewater purification process. Under the author supervision, Senik E.V. conducted detailed research in this direction [26].

In case if flotodecanter with first defecating chamber is chosen, its type choice (column or with consequent chambers) is made in dependence of the space for equipment installation. If the area for flotoharvester installation is small, we choose flotodecanter of column type as its advantage is a compactness. If it is enough space for equipment installation, it is necessary to choose the flotodecanter with consequent chambers [16].

Aeration system, working depths of defecating and flotation zones, fin block parameters (width B, minimal distance between fins b_{def} (from defecating chamber side)), speeds in the chambers of defecating v_{def} and flotation v_{fl} are chosen on that stage.

Let accept the speed in defecating chamber v_{def} from 5 to 10 mm/s, speed in flotation chamber v_{fl} — no more than 5 mm/s, the distance between fins of fine-zoned defecating block with diverging fins from defecating chamber side b_{def} is accepted from 12.5 до 141 mm.

According to Senik E.V. data, the efficiency of one element of fin block is:

$$q = B \cdot b_{def} \cdot v_{def}.$$

Let's find the minimal quantity of the elements of block of fins n that is necessary to provide for set expenditure of being purified wastewaters:

$$n = \frac{Q}{q},$$

where Q — wastewater expenditure.

Let's round the obtained value n till integer. Then the number of fins in block is $n + 1$. Let's set inclination angle values of lower fins of block lower and upper elements as α_1 и α_2 (in the case if the calculation of block with diverged fins is held which promotes particle sedimentation, the fin inclination angles are recommended to be accepted equal from 40 to 60°). Let's find maximal angle between fins β:

$$\beta = \frac{\alpha_2 - \alpha_1}{n}.$$

Let's check if found angle is less than the angle, denoted in Table 5 for set parameters.

If fin block fulfills only matching of hydrodynamic regimes, the calculation is pursued by the following way. Distance between fin-block fins from the flotation chamber side b_{fl} is calculated:

$$b_{fl} = \frac{v_{def} \cdot b_{def}}{v_{fl}},$$

where b_{def} и b_{fl} — distance between fins from defecating and flotation chamber sides, correspondingly, m.

Let's find fin length, necessary for hydrodynamic regimes matching, according to the formula:

$$L_{match} = \frac{b_{fl} - b_{def}}{2\sin\dfrac{\beta}{2}}.$$

Let's then calculate time-being of effluent, being purified, in fin block element:

$$t_{match} = \frac{L\left(b_{def} + fl\right)}{2b_1 \cdot v_{def}}.$$

If block of diverging fins has to provide for the sedimentation of particles of hydrodynamic size v, besides hydrodynamic regime matching in defecating chamber, then

$$\left\{ \begin{array}{l} t_{sed} = \dfrac{r_0}{2v_{sed}}\left[\left(1 - \dfrac{v_{sed}}{u}\left[\ln\dfrac{\left(tg\left(\dfrac{\pi}{4} + \dfrac{\alpha}{2}\right)\right)}{tg\left(\dfrac{\pi}{4} + \dfrac{\alpha+\beta}{2}\right)}\right]\right)^2 - 1\right]; \\[2em] r_{sed} = \sqrt{2r_0 v_{def}t + r_0^2} \end{array} \right.$$

88

where $r_0 = \dfrac{b_{def}}{2\sin\dfrac{\beta}{2}}$.

Fin length necessary for sedimentation $L = r_{sed} - r_0$.

If $L_{match} > L_{sed}$, then $L = L_{match}$, $\tau = \tau_{match}$ and other parameters are accepted according to the calculation for regime matching.

If $L_{match} < L_{sed}$, then $L = L_{sed}$, $\tau = \tau_{sed}$. Let's find additionally the corresponding values for distances between fins from the side of flotation chamber b_{fl} and the average speed from the side of flotation chamber v_{fl}:

$$b_{fl} = 2L\sin\frac{\beta}{2} + b_{def};$$

$$v_{fl} = \frac{v_{def} \cdot b_{def}}{b_{fl}}.$$

Calculation of time of wastewater purification in flotodecanter and its working volume for various types is held depending on flotodecanter type.

There calculated separately the times of being in preparation zones of being purified effluent (τ_1), of separation in block of diverging fins (τ_3), in defecating zone (τ_2) and in flotation zone (τ_4) (Fig. 5.3).

Time τ_1 necessary for wastewater mixing with reagents, is accepted equal from 3 to 5 min that is analogously to time-being of effluent in hydraulic blender. Time, necessary for flake-formation is accepted from 6 to 30 min depending on flake-formation chamber type. Overall time of effluent being in flotation zone τ_4 is calculated according to flotation multistage model of Ksenofontov B.S. Initial data for the calculation: bubble average diameter D, micron, barbotage speed q, m³/(m²·s); efficiency of pollution particle capture by surface-emerging bubble E; flotocomplex surface-emerge speed v_{em}, mm/s; flotation chamber height h_f.

In the case of well-floating pollutions, one can use a simplified model.

If we can't ignore reversibility, then we use flotation general model. Thus, we additionally calculate constants k_4, k_5, k_6, k_7.

Then, we solve system of differential equation, describing flotation at initial conditions $t = 0$ $C_A(0) = C_0$, $C_B(0) = C_C(0) = 0$. The graph of change for pollution concentrations in flotation chamber is presented on Fig 5.4.

Let determine flotation time τ_4 from the graph for set purification efficiency.

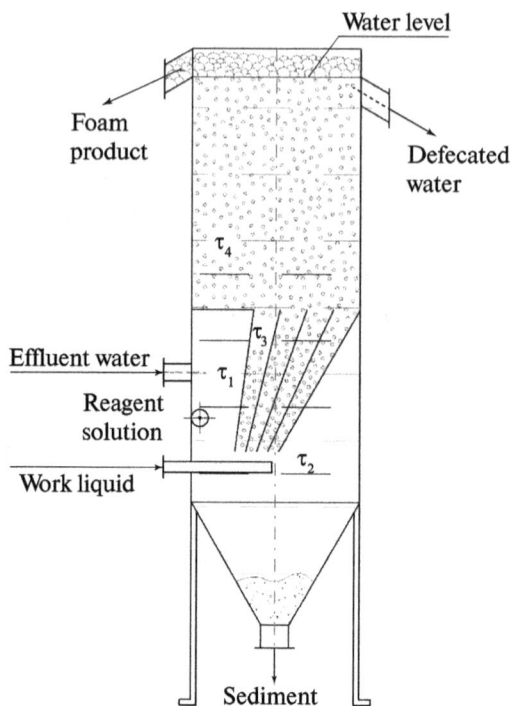

Fig. 5.3. Scheme of flotodecanter of column type:
τ_1 — preparation time of being purified effluent; τ_2 —defecating time;
τ_3 — separation time in block; τ_4 — flotation time

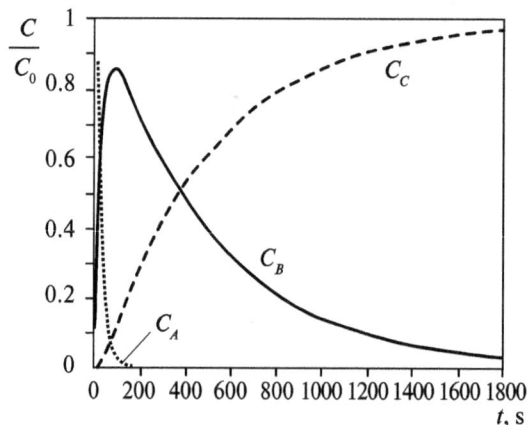

Fig. 5.4. Graph of change of pollution concentrations in flotation chamber

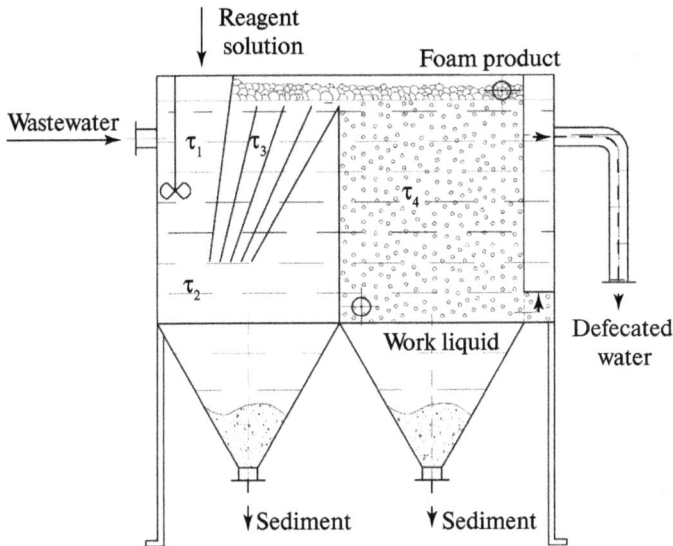

Fig. 5.5. Scheme of flotodecanter with consequent chambers

Knowing purification time in each zone, we can find overall purification time τ by the way of summarizing of purification times of each stage. Besides, knowing wastewater expenditure Q, one can find working volumes of all purification zones, and overall working volume of the apparatus:

$$V_{work} = Q \cdot \tau.$$

In the case of flotodecanter with consequent chambers, it is separately calculated time-being in the zones of preparation of being purified effluent (τ_1), time of separation in block of diverging fins (τ_3), in defecating zone (τ_2) and flotation zone (τ_4) (Fig. 5.5).

Time-being in the chamber of being purified effluent mixing with reagents is accepted equal from 0.5 to 3 min that is analogous to mechanical blenders. Time, necessary for flake-formation, is accepted to be from 6 to 30 min depending on flake-formation chamber type. Time-being of effluent in preparation zone (τ_1) is found by the way of mixing and flake-formation times summarizing.

The calculation of purification time in flotodefecating chamber is held upon the model of purification for the effluent, containing hydrophobic and hydrophilic pollutions, by flotation-gravitational approach which is afore-described in the manuscript.

Initial data for the calculation: bubble average diameter D, micron, barbotage speed q, m^3/(m$^2 \cdot$s); efficiency of pollution particle capture by sur-

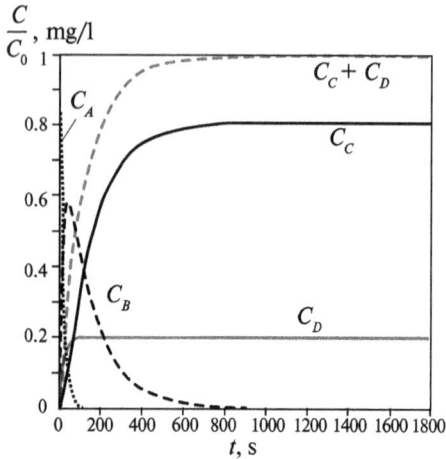

Fig. 5.6. Graph of change of pollution concentrations in flotodefecating chamber

face-emerging bubble E; floto-complex surface-emerge speed v_{em}, mm/s, sedimentation speed v_{sed}, mm/s; flotation chamber height h_f; defecating chamber height h_{sed}, m.

Then, we solve the system of differential equations, describing purification of effluent, containing hydrophobic and hydrophilic pollutions, by flotation-gravitational approach at initial conditions $t = 0$, $C_A(0) = C_0$, $C_B(0) = C_C(0) = C_D(0) = 0$. Graph of pollution concentration change in flotodefecating chamber versus time is shown on Fig. 5.6.

On the graph basis, let choose purification time in flotodefecating chamber depending on the necessary efficiency and let find purification overall time and apparatus working volume.

5.3 WASTEWATER PURIFICATION INTENSIFICATION WITH THE USE OF COMBINED FLOTATION TECHNIQUES

The application of combined flotation apparatuses, including floto-harvesters, in wastewater purification practice has shown the effectiveness of their usage [29, 50].

For instance, in the row of electro-stations with heating and condensational turbo-devices, the problem of getting of turbo oil in major condensate line can take place. It proceeds in the connection with the leakage of oil along turbine shaft via chamber of seals and stream ejectors into ejector cooler and further into major condensate system. The pointed oil leakages are forced at non-stationary regimes, for instance, at starts and stops of turbo-devices. Usually, at steady burden, oil minerals content in seal ejectors' condensate doesn't exceed 5 mg/l. At starts and stops of turbine, mineral oil concentration can grow till 50 mg/l and more that leads to significant decrease of the resources of ion-exchange resins of filters for block desalted device (BDD).

In this regard, the most rational settlement of this problem is the local purification of seal ejector condensate from mineral oils. Whereat, purification

92

technology, if possible, shouldn't include pressure-head filtering elements, as it will significantly complicate turbo-device exploitation. It is especially important to pursue the purification from mineral oils of unit discharges. For to settle the pointed task, mineral oil traps of defecating type are installed on energetic blocks of thermo-electro-station "Yuzhnaya" of "Lenenergo" Co, equipped with turbo devices T-250. Exploitation experience of these mineral oil traps has shown that the problem of condensate purification cannot be solved with their help. There are two causes: condensate heightened temperature (55–70 °C) and water flow high speed. For to provide seal condensate cleaning we offered the pneumatic type flotation machine which test sample is installed on energy block № 2. Pneumatic flotation machine was installed on the stage of fine purification of technological condensate after the existing mineral oil trap.

Pneumatic flotation machine (Fig. 5.7) includes Shell 2, divided into four Chambers 4 with, installed in lower part, pored Aerators 12. Aerators can be accomplished from deformed (for instance, rubber or polyethylene) or non-deformed (for instance, ceramics or metal) material. In test sample of pneumatic flotation machine, installed on energy block № 2, the low-stable rubber with the size of special holes (pores) of 0.5 mm was used as a material for aerators.

In additional Chamber 5, the special Block of fine-zoned defecating 6 as well as Device of liquid level regulation 9 are installed. On external side of flotomachine corps, the entry Sleeve 1 and exit Sleeve 10, respectively, for input and output of wastewater as well as foam Gutter 7 with exit Sleeve 13 for output of caught pollutions in the form of emulsion are installed.

Working principle of pneumatic flotation machine is in the following. Original (dirty) water is supplied via entry Sleeve 1 and, further, it moves in horizontal direction via Chambers 4 with pored Aerators 12 and additional Chamber 5 with fine-zoned defecating and is withdrawn from the machine via level regulation Device 9 and exit Sleeve 10.

During the water movement through first four chambers, its aeration (barbotage) proceeds by bubbles of air, introduced under pressure via pored aerators. Whereat, air bubbles stick with hydrophobic pollutions and surface-emerge in the form of flotocomplexes (mineral oil drops)-(air bubbles). Caught pollutions in the form of foam product (emulsion with mineral oils concentration of 10–20%) are removed by drift via Gutter 7 and exit Sleeve 13.

It is known that flotomachine work effectiveness much depends on the type and design of aerators. In this regard, it should be considered in more details the design and action principle of applied aerators in these machines.

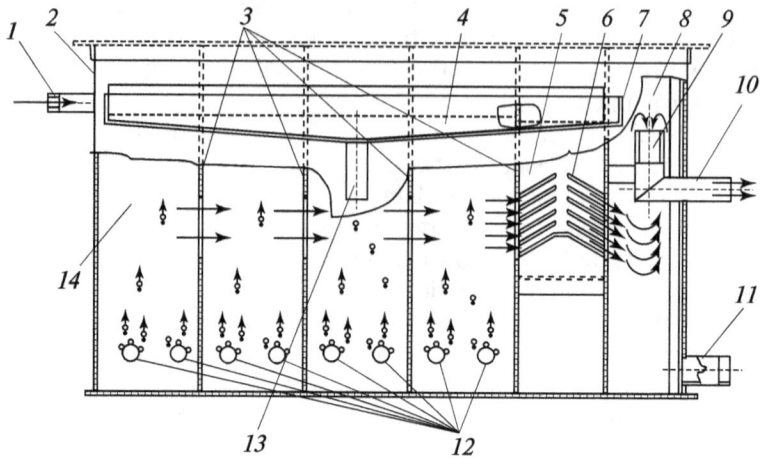

Fig. 5.7. Scheme of pneumatic flotation machine
of SFM-0.5 type

In pneumatic type machines, the tubular aerators are used predominantly; these aerators usually constitute welded design from tubes; the design is presented by framework with central collector which leading and muffled sleeves are installed on. Whereat, chokes are evenly spread on the framework; dispersing elements from perforated rubber tubes are installed and fixed by clamps on the chokes.

Pneumatic flotomachine trials were held in two stages. On the first stage, aeration regime was treated mainly, this regime essentially effects the effectiveness of flotation purification condensate. Table 5.1 shows results of the condensate purification depending on the aeration intensity at water level in flotation skimmer of 0.9 m.

Data analysis, presented in Table 5.1, shows that the highest effect of condensate purification by flotation is reached at aeration intensity 0.8 m^3 on 1 m^2 condensate surface in a minute. While the pursuing of concluding trials in the same conditions, it was established that mineral oil concentration in purified water, going out from flotomachine, is 0.3–2.0 mg/l (sometimes till 4–5 mg/l).

Pointed quality of technological condensate purification was confirmed as a result of pursuing of long-term testing for pneumatic flotation machine sample in Yuzhnaya TES (thermo electro-station) BTW (boiler-turbine workshop) on energy block n.2. Trial total results are shown in Table 5.2.

Table 5.1. Dependence of condensate flotation purification effectiveness versus intensity of condensate aeration by air

Sample number	Aeration intensity, $m^3/m^2 min$	Mineral oil concentration, mg/l		Purification effect, %
		Before purification	After purification	
1	0.3	7.2	4.1	43
2	0.4	7.2	3.9	46
3	0.5	7.2	2.3	68
4	0.6	7.2	1.8	75
5	0.7	7.2	1.1	84
6	0.8	7.2	0.7	90
7	0.9	7.2	1.3	82
8	1.0	7.2	1.9	74
9	1.1	7.2	2.7	63
10	1.2	7.2	3.8	47

Table 5.2. Overall results of industrial testing of pneumatic flotation machine SFM-0.5 on energy block n.2 Yuzhnaya TES of "Lenenergo" Co

Sample number	Mineral oil concentration in technological concentrate (mg/l)	
	Before purification	After purification
1	2.34	0.36
2	41.0	3.7
3	37.0	3.2
4	51.3	4.2
5	28.6	2.0
6	10.1	0.4
7	4.7	0.7
8	8.4	1.0
9	2.6	0.3
10	4.1	0.5

Pneumatic flotation machine data testing, presented in Table 5.2, shows that the quality of water purification from mineral oils doesn't exceed 4–5 mg/l even at large initial concentrations of mineral oils in initial condensate. This allows to guarantee pointed quality at similar installation exploitation on the stage of purification of technological condensate from mineral oils.

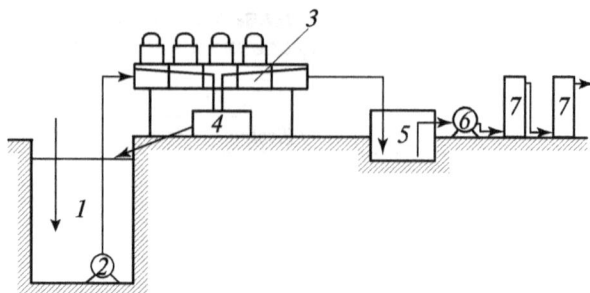

Fig. 5.8. Scheme of surface wastewater purification

The purification of surface wastewaters with the achievement of standard quality of water purification is very important practically on all energy enterprises.

Let consider in more details the surface wastewater purification scheme, offered by us. Surface wastewaters from industrial territory, going through lattice, gather in volume-decanter (Fig. 5.8).

Wastewater from Volume *1* is pumped out by Pump *2* and is supplied into pneumatic flotation Machine *3* of SFM-0.5 type with fine-zoned block of defecating. In the mentioned flotation machine (Fig. 5.7), fine-dispersed extraction of mineral oil products takes place at their surface-emergence together with air bubbles, formed during air dispersal by the way of its supply under pressure via pored aerators, made of special rubber. Aerators in the quantity of *12* are based by *3* in each of four chambers of the pointed flotation machine. In additional fifth chamber of the flotation machine, the block fine-zoned defecating is placed for fine-extraction of fine-dispersed drops of mineral oils. Being purified wastewater consequently goes through all pointed chambers, whereat, caught mineral oil pollutions in the form of foam product are collected in the upper part of being purified water layer. Surface-emerging mineral oil products together with air bubbles create the foam layer which is by drift removed into foam product Collector *4* (pack-delivery with the flotation machine). After defecating of the foam product, presenting itself a mixture of water and caught mineral oils, the decanted water is leaked out into deepened Volume *1*. The purified liquid is withdrawn from the flotation machine by the way of consequent passing via fine-zoned block and via device of being purified liquid set level maintenance in the flotation machine and is supplied by drift into intermediate Reservoir (collector) *5* with working volume of no less than 3 m³. The intermediate Reservoir *5* is made of monolithic or fabricated ferroconcrete.

With the help of surface Pump 6, the preliminary purified water is supplied for fine-purification to sorption Filters 7. The first by liquid flow filter has combined load, including expanded clay layer (lower layer) and activated carbon layer, and the second filter is fully loaded with activated carbon. The carbon close to carbon brand AG-3 is used in these filters of pressure-head type. It allows to pursue deep fine-purification of surface wastewaters till residual oil mineral content of no more than 0.05 mg/l. Considering the seasonal specifics of purification device working, it is offered not to regenerate the carbon load but to use it only during one season. Worked-out load is feasibly to remove by burning, for instance, in boiler room or in special oven, where coal is used as a fuel. Purified wastewater with mineral oil content of no more than 0.05 mg/l can be thrown on relief or into nearby pond.

The discussed type flotomachine has been introduced in many enterprises and, in this regard, technological scheme of wastewater purification in particular cases has differed from the above-described one. Wastewater purification devices, including flotomachine, which scheme is presented on Fig. 5.5, have been introduced in various energetics enterprises. Table 5.3 shows results of control (concluding) trials of above-described device during purification of surface effluent of auto-transport workshop of Ust'-Ilyim Hydro-ElectroStation.

Testing results, presented in Table 5.3, testify the achievement of normative purification quality of surface wastewaters while using of above-described technology.

The essential improvement of SFM-type (surface flotation machine) machines is a flotation machine with filtering element. This flotation machine for wastewater purification includes shell (Fig. 5.12) which is divided by walls into chambers with aerators, installed inside them near bottom, and fine-zoned defecating block, situated in the last chambers forwards, and device for waste liquid level regulation, which is on the external side of shell of entry and exit sleeves, and whereat, the distinguished specificity of offered flotomachine is additionally fixed intermediate chamber on the side of gutter for caught pollutions bending. The chamber is between fine-zoned defecating block and liquid level regulation device, with filtering element therein; the element is accomplished in the form of two empty cylinders with perforated surfaces, with load between them, whereat the interval of live section values of perforated surface has been checked as a result of pursued experimental research and constitutes 10–25% of overall space. Meanwhile, it has been established that at live section of less than 10% the increase of wastewater purification efficiency is not reached, but, at live section of more than 25%, the reached effect doesn't change.

Table 5.3. Results of control (concluding) testing of surface effluent of auto-transport workshop of Ust'-Ilyim HES, "Irkutskenergo" Co.

Defined indicator	ND (normative documentation) for FIM (fulfilled instruction methodology)	Measurement result mg/dm³		Measurement result deviation mg/dm³		Note
		Decanter	after filters	Decanter	after filters	
Suspended substances	NNDF (nature-protecting normative documentation of Federal level) 14.1:2.110–97	114.6	11.3	for all	for all	norm 0.75 to background
		120.1	11.1	10.0	2.0	
		112.4	11.2			
		116.6	10.8			
		117.8	10.5			
Mineral oils	NNDF 14:1:2:4.128–98 Edition, 2002.	4.99	0.02	1.25	0.01	
		1.48	0.02	0.37 0.36	0.01	
		1.42	0.03	0.37	0.015	
		1.49	0.04	0.29	0.02	
		1.16	0.02		0.02	

The load is fulfilled with adsorption material, for example, active carbon, sipron, visopron, megasorb and so on.

Flotation machine with the filtering element for wastewater cleaning consists of Shell *1* (Fig. *5.9*), which external side Sleeve *2* for the supply of original wastewater, foam Gutter *3* and Sleeve *4* for pollution withdrawal in the form of foam product, Sleeve *5* for purified water withdrawal are installed on. Inside the shell, there are installed disk Aerators *6*, which gas (air) is supplied via, as well as half-deepened Walls *7*, fine-zoned defecating Block *8*, Device *10* for liquid level regulation and filtering element, comprising Framework *10* with perforated external and internal surfaces *11*, which filtering Load *12* is situated between.

Work principle of exploited flotation machine for wastewater purification is in the following. Initial wastewater is supplied via Sleeve *3* and then into lower part of flotomachine Shell *1* into aeration zone, created by disk Aerators *6* which are generators of bubbles of gas (air). During the movement of water in aeration zone, the hydrophobic dirt sticks into particle-bubble complexes upon contact with bubbles. Formed flotocomplexes, pollution particles and air bubbles lift upwards, forming foam layer which is being withdrawn from

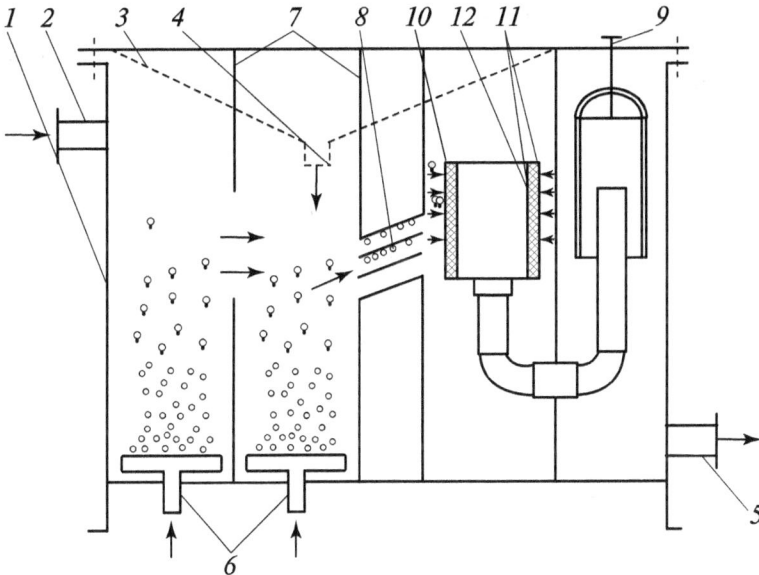

Fig. 5.9. Scheme of flotation machine with filtering element

machine corps by drift or forcibly via foam Gutter *3* and Sleeve *4* into mud collector. Purified water from aeration zone is being withdrawn via Block *8* of fine-zoned capture of bubbles. Flotocomplexes, got into between-shelves space, quickly reach the upper shelf because of the small distance between shelves (from 10 to 150 mm). Clung to upper shelves, flotocomplexes combine into larger aggregates that provides for large expelling force occurrence and quick surface-emerging of these complexes into upper foam layer. Cleaned liquid, gone through between-shelves space, gets into filtration zone where filtering element of adsorption type is present. Clung flotocomplexes on external perforated Surface *11* of filtering element coalesce between each other analogously as in fine-zoned defecating Block. Inside filtering element, there is special Material *12*, catching dirt being in suspended state. Then, purified water goes through Device *9* of liquid level regulation and is withdrawn from flotomachine shell via Sleeve *5*. Wastewater purification efficiency in offered flotation machine constitutes 98−99.5 %, meanwhile, absolute value of hydrophobic pollutions in being purified liquid constitutes from about 0.05 to 0.5 mg/l that is higher in efficiency by far than in the known flotation machines wherein hydrophobic pollution residual concentration is no less than 2−4 mg/l as a rule.

In individual cases, the reaching of quality normative indicators of water purification according to described above technologies seems complicated, in particular, because of the presence of hydrophilic pollutions significant number. In this case, the application of coagulants and flocculants can be effective. Water purification efficiency raise depends on their application respectively coagulant and flocculant properties, method of their mixing with water, dirt nature and so on.

Improved samples of flotation technique, including SFM–0.5. were included into various projects of reconstruction of existing technological installations and newly created purification buildings, in particular, on energetic complex enterprises.

One of the variants of modernization of already existing pneumatic flotomachines might be the usage on exit of filtering elements on the basis of non-woven materials. Testing of such materials with their location in machine exit chamber was pursued on Ust'-Ilyim HydroElectroStation "Irkutskenergo" Co (Fig. 5.10).

Measurements have shown that while usage of such combined flotomachine, the purification of oil-containing effluents from 0.79–0.54 mg/l has been held, and while usage of complex flotomachine-(adsorption filters), the oil residual concentration in purified water has constituted less than 0.05 mg/l.

Major canon for filtering load choice was material the least resistance. Perforated material, which filtering element corps is made of, was calculated respectively the sizes of flotocomplexes and hydrophilic pollutions in the way to provide for their coalescence.

Fig. 5.10. *a* — Common view of combined flotation machine;
b — Exit chamber of flotation machine with filtering element

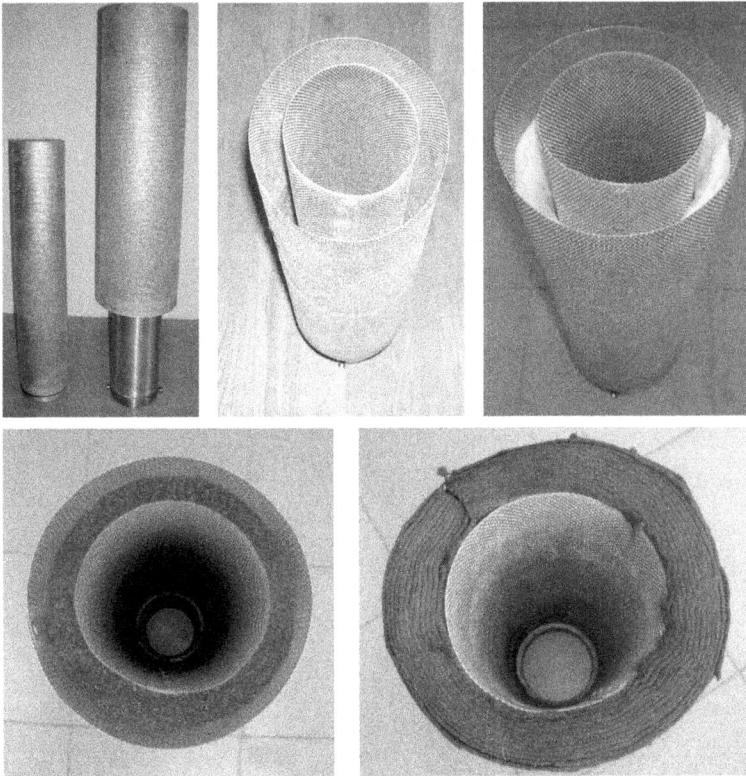

Fig. 5.11. Filtering elements for combined flotation machine

Pursued trials of filtering elements with load various kinds (Fig. 5.11) have shown that the most effective filtering is one through the material with carbon fibers UVIS-AK-V which allows to reach wastewater purification indicators to the demands met at dumping into an open pond. An application of material "Megasorb" or carbon load is possible only at the use of wastewater purification technology with recursive water usage application.

5.4 WASTEWATER PURIFICATION IN FLOTATION COLUMNS

Wastewater flotation purification in much extent depends on dispersed air quantity and formed gas bubble size as well as on the conditions of gas bubbles contacting with pollution particles, predominantly used in the practice of wastewater cleaning.

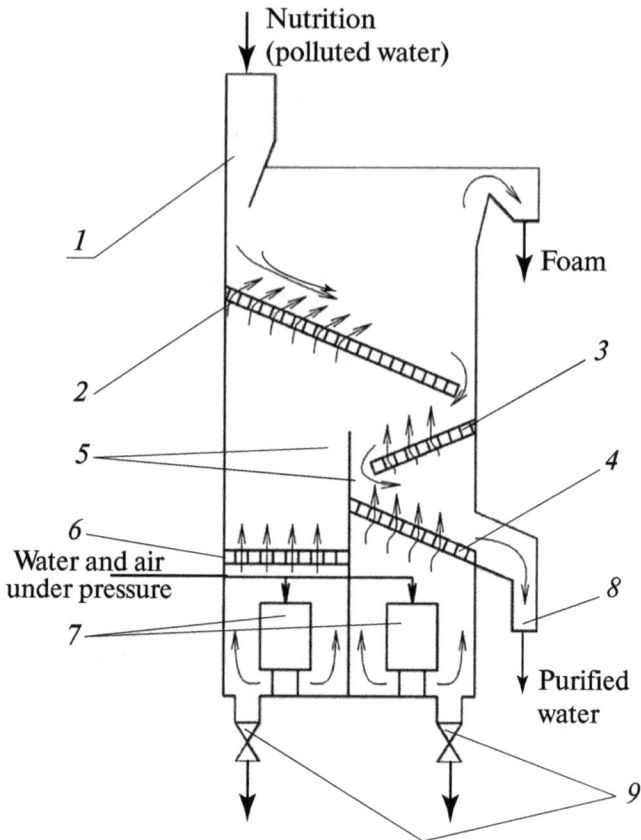

Fig. 5.12. Flotation column apparatus
with inclined and dispersed lattices:
1 — water supply; *2, 3, 4* — inclined lattices;
5 — first and second chambers; *6* — dispersed lattice;
7 — jet aerators; *8* — water bend; *9* — water emergency drop

In this regard, in the row with earlier worked-out and introduced in the series of enterprises by mechanical and pneumatic flotomachines, we were testing several new types of flotation machines and apparatuses wherein contacting conditions of gas bubbles with pollution particles proceed in more favorable conditions than in typical flotation machines and apparatuses, for instance, in flotation column apparatus with inclined and dispersed lattices (Fig. 5.12).

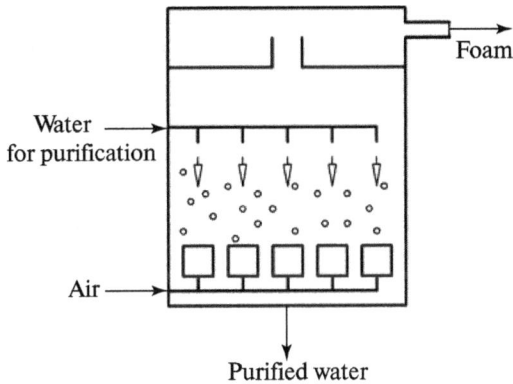

Fig. 5.13. Laboratory flotocolumn scheme

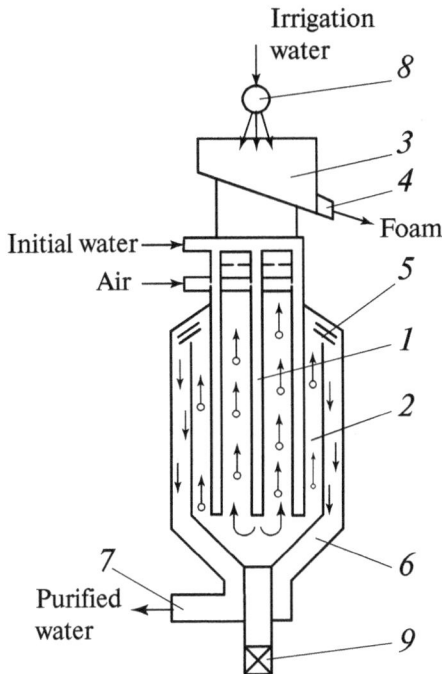

Fig. 5.14. Experimental-industrial flotocolumn sample:
1 — deepening jet aerators; *2* — flotation chamber; *3* — reservoir
for foam; *4* — foam gutter; *5* — fin-catcher of bubbles;
6 — water bend cell; *7* — bend channel; *8* — irrigator;
9 — water emergency bend

Rather high probability of particles collisions with bubbles thanks to meeting movement of particles and bubbles, high probability of conglutination and safety of mineralized air bubbles is provided in flotation columns. In the columns, relative speeds of bubbles and dirt particles alternate in the limits 10−12 cm/s that according to experimental data, creates optimal conditions for their conglutination. Mixing devices, creating inertia forces, causing particle detachment from bubbles, are absent in the columns that heightens complexes (pollution particle)-(bubble) safety.

As advantages of counter-current columns there should be stated also low energy-capacity, small capital costs, small space, necessary for installation, constituting approximately 20−30% of space occupied by standard machines of the same productivity as well as wide possibility of usage of secondary mineralization processes in foam layer for the increase of efficiency of conglutination between bubbles and dirt particles.

To confirm the pointed advantages the testing has been pursued with the use of laboratory (Fig. 5.13) and experimental-industrial (Fig. 5.14) samples.

Testing of efficiency of the use of flotation column apparatuses was held while cleaning of oil-containing wastewaters. In this regard, on the stage of laboratory research, experimental sample of flotation column with working volume 15 l was used, and on the stage of experimental-industrial tests — sample of flotocolumn with working column of 1 m³ volume. Results of laboratory investigations are shown in Table 5.4.

Analysis of data, presented in Table 5.4, testifies the achievement of rather low concentration of mineral oils in purified water not exceeding 1.8 mg/l during time of oil effluent flotation of no more than 22.5 minutes. Such indicators in wastewater purification practice are considered to be high, and they point on perspectivity of the use of flotation column apparatuses for wastewater cleaning from hydrophobic pollutions, for instance, from mineral oils.

For to check the efficiency of the usage of flotation column technique in experimental-industrial conditions, the trials on apparatus with working volume of 1 m³ have been held. The results of these trials are presented in Table 5.5.

Data, presented in obtained earlier Table 5.5, testify the confirmation of trial results obtained earlier in laboratory conditions. Some difference, in our opinion, in the values of residual concentration of oils in purified water, moving out from laboratory and experimental-industrial samples of flotation machine, is due to largescale difference in pointed apparatuses.

Table 5.4. Flotation time influence on oil residual concentration in purified water (surface wastewaters with oil concentration in original water 27.8 mg/l; wastewater aeration intensity — 1.1 m³/m² min)*

Order number	Flotation time, min	Oils concentration in purified water, mg/l	Purification effectiveness (%)
1	5	15.6	43.9
2	7.5	11.3	59.4
3	10	7.9	71.6
4	12.5	5.1	81.6
5	15	3.7	86.7
6	17.5	2.2	92.1
7	20	1.9	93.2
8	22.5	1.8	93.5
9	25	1.8	93.5
10	30	1.8	93.5

*Tests were held in periodical regime.

Table 5.5. Averaged values of oil concentration in purified water depending on wastewater flotation time in flotation column with working volume of 1 m³*

n.	Flotation time, min	Oil concentration in original water, mg/l	Residual concentration of oils in purified water, mg/l	Purification effectiveness (%)
1	5.7	32.4	16.6	48.8
2	10.1	30.1	13.1	56.5
3	14.5	28.9	6.9	76.1
4	20.2	31.6	4.8	84.8
5	23.9	29.8	3.3	88.9
6	30.6	31.2	3.4	89.1

*Trials were held in continuous regime

Obtained data are easy to interpret also from multistage flotation theory. The decrease in gas bubble sizes leads to the increase of constant, characterizing the probability of pollution particle and bubble adhesion. This constant value grows also at the increase of probability of pollution particle capture by gas bubble that is also observed in column flotation apparatuses, working in reverse-flow regime, i.e. at meeting movement of

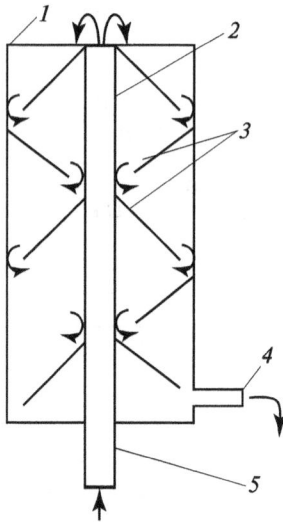

liquid (water) and gas phases. It's worth to underline that the probability of dirt particle capture in column flotation apparatuses raises, as the results of our special studies show, approximately by 1.5 times in comparison with this parameter value in typical flotation machines. Namely marked circumstances, first of all, appear in the majority of cases as determining ones at choice of flotation apparatus type for wastewater purification from hydrophobic dirt.

Fig. 5.15. Scheme of membranous apparatus for water saturation with oxygen:

1 — apparatus corps; *2* — tube for original (dirty) water supply; *3* — inclined shelves; *4* — exit sleeve; *5* — entrance sleeve

Together with cleaning, it is important also the wastewater conditioning, in particular, their saturation with oxygen. As the results of our research have shown, this technological operation can be made in column membranous apparatuses.

Membranous apparatus for water saturation with oxygen in thin layer includes cylindrical Shell *1*, which Tube *2* is aligned inside with exit Sleeve *5*, and Shelves *3* are between Tube and Shell. Withdrawal of water, saturated with oxygen, is made via Sleeve *4*.

Comparing trials of flotation column with jet and ejector aerators and of membranous apparatus, differing by the highest effect of water saturation with oxygen, were held at the same temperature regimes ($T = 22\,°C$).

Data on holding of trials of membranous and column flotation apparatuses are written in Table 5.6 ($T = 22°C$, $P = 1.5$ atmospheres; theoretically possible saturation with oxygen at given temperature is 8.83 mg/l).

The data given in Table 5.6 show that water saturation with oxygen proceeds the most efficiently in flotation column apparatus with aeration combined system. These data give substantiation of column flotation apparatuses usage not only for wastewater purification but also for wastewater saturation with oxygen before dropping into an opened pond.

*Table 5.6. Comparison of efficiencies of water saturation
with oxygen with the use of flotation column with jet
and ejector aerators and of membranous apparatus*

n	Oxygen initial concentration, 10^{-3} kg/m^3	Oxygen ending concentration, 10^{-3} kg/m^3
	Membranous (fine-zoned) apparatus	
1	5.1	7.1
2	5.0	7.0
3	5.2	7.1
	Flotocolumn with ejectors	
4	5.0	7.7
5	5.7	7.8
6	6.8	8.1
	Flotocolumn with jet aerators	
7	5.5	7.8
8	6.8	8.0
9	7.0	8.05

5.5 THE USE OF JET AND EJECTION AERATORS FOR WASTEWATER PURIFICATION

It's known that the efficiency of wastewater flotation purification significantly depends on used aeration system [36–47]. It was shown earlier that application of jet and ejector aeration systems in the series of cases leads to the reaching of high efficiency of wastewater purification. In this regard, joint use of jet and ejector aeration systems seems feasible, for instance, with the use of flotation columns.

*Table 5.7. Averaged values of oil concentration in purified
water depending on wastewater flotation time
in flotation column of working volume of 0.5 m^3*

n	Flotation time, min	Oil concentrations in original water, mg/l	Residual concentration of oils in cleaned water, mg/l	Purification efficiency (%)
1	5	22.5	12.4	48.8
2	10	20.7	10.9	47.3
3	15	18.8	5.8	69.1
4	20	21.8	4.6	78.9
5	25	19.4	2.2	88.6
6	30	20.1	2.5	87.6

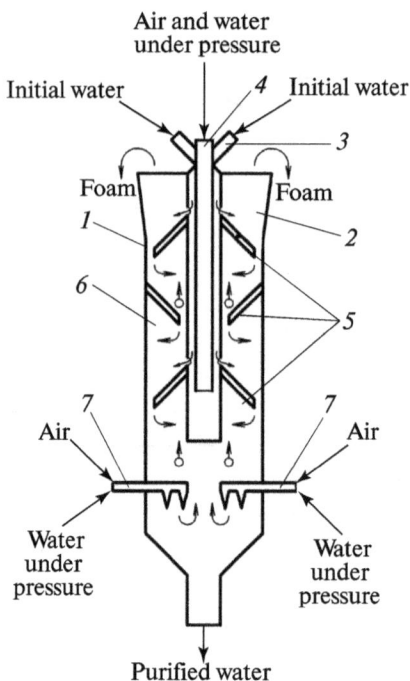

Fig. 5.16. Flotation column with jet aerators and ejectors:

1 — column; *2* — reservoir for foam; *3* — water supply; *4* — jet aerators; *5* — inclined shelves; *6* — jet aeration chamber; *7* — ejectors

To check the efficiency of column flotation technique usage in experimental-industrial conditions, the testing was held on the apparatus with work volume $0.5\ \text{m}^3$ (Fig. 5.16).

These trials results are shown in Table 5.7.

The data given in Table 5.7 testify high efficiency oil containing wastewater cleaning.

It is easy to interpret the obtained data also from the side of multistage flotation theory, being developed by us during last 30 years [15, 30]. Observed gas bubble size decrease in these cases leads to the increase of constant, characterizing sticking probability of dirt particle with bubble. This constant value increases also at the raise of probability of gas bubble capture of dirt particle that is observed in column flotation apparatuses, working in reverse-flow regime, i.e. at meeting movement of liquid (water) and gas phases. It's worth to note that the probability of dirt particle's capture in column flotation apparatuses increases, as the results of our special research show, approximately 1.5 times in comparison with this parameter values in typical flotation machines. Namely noted circumstances, first of all, appear to be in the majority of cases determining ones in flotation apparatus choice for wastewater purification from hydrophobic pollutions.

The confirmation of said suppositions was checked in laboratory conditions by stages, beginning from aeration process study by freely falling jet.

Comparative trials of purified waters aeration with the use of freely falling jet and submerged jet have shown that the effectiveness of water aeration with the application of freely falling jet is significantly (about 2–3 times) higher than of submerged jet. Whereat, aeration effectiveness was determined by known approach by the definition of water displacement from flask, preliminary filled with water and then turned down with neck downwards with the immersion into being aerated water. In this case, the less the water displacement time from flask the more intensive aeration is, including oxygen solubility in water. Aeration scheme of freely falling jet is presented on Fig. 5.17. The research of the influence of jet fall height and angle on aeration intensity was pursued.

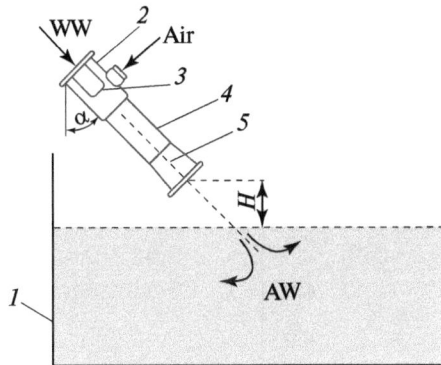

Fig. 5.17. Scheme of aeration by freely falling jet:
1 — reservoir with water; 2 — ejector; 3 — nozzle; 4 — chamber of ejector mixing; 5 — diffusor; α — jet fall angle; H — height of ejector diffusor location above water level; WW — wastewater; AW — aerated water

For to pursue the given research the installation, comprising ejector with prolonged conic chamber of mixing, was used.

During the research pursuing process, it was established that, in the case of freely falling jet, the value of jet fall angle and the height of diffusor, generating jet flow, have large significance. The obtained dependencies are shown on graphs (Fig. 5.18–5.19).

Despite that effectiveness of aeration by freely falling jet is higher, the aeration by submerged jet simplifies apparatus decoration and technological usage of ejectors as aerators. Therefore, the investigations on influence of jet submergence depth and air expenditure on aeration intensity have been pursued.

Trials on the defining of air expenditure for aeration intensity were held on the device, presented on Fig. 5.21.

To define air expenditure influence for aeration intensity, the trials were pursued with the maintenance of water constant expenditure of 0.14, 0.18 and 0.22 1/s. In all 3 cases, with air expenditure raise, the aeration intensity increased. The obtained data are demonstrated in the form of graphs (Fig. 5.22, a, b, c).

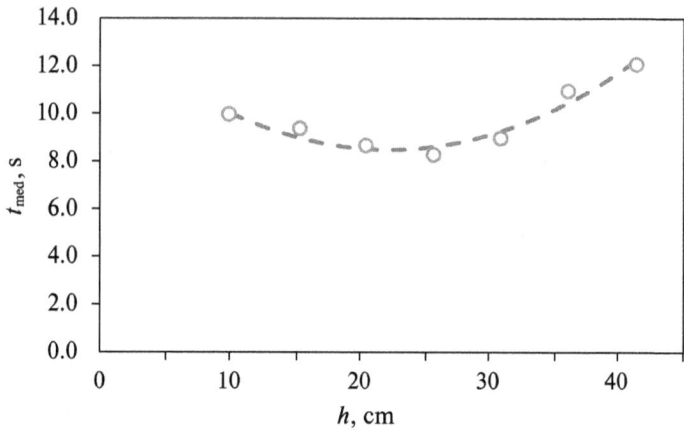

Fig. 5.18. Aeration intensity dependence from height h above liquid level at fall angle value 300 of diffusor, generating jet

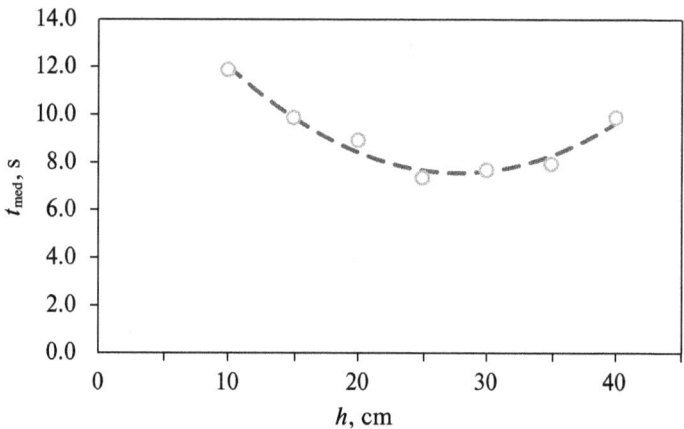

Fig. 5.19. Aeration intensity dependence from height, above liquid level, of diffusor, generating jet, at fall angle value 45°. As a result of pursued research, it was established that aeration intensity is maximal at fall angle of 45° and diffusor height above liquid level in interval about 20–35 cm at freely falling jet

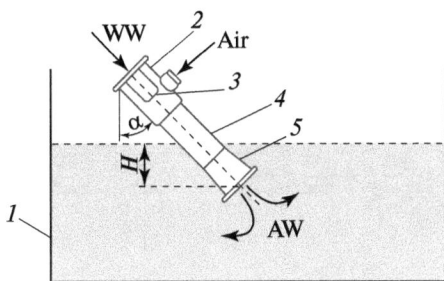

Fig. 5.20. Scheme of aeration
by submerged jet:
1 — reservoir with water;
2 — ejector; 3 — nozzle;
4 — mixing chamber of ejector;
5 — diffusor; α — jet fall angle;
WW — wastewater;
AW — aerated water

Thus, it has been established as a result of research that aeration intensity is maximal at fall angle of 45° and diffusor height above liquid level in interval about 20–35 cm while freely falling jet. Water aeration intensity with the use of freely falling jet is significantly higher (about 2–3 times) that of submerged jet.

Depending on wastewater volume, ejector quantity can change; it's worth to use 2 (Fig. 5.23a) or 4 (Fig. 5.23a) ejectors for the apparatus of low productivity.

One of the drawbacks of ejector usage is predominantly a local aeration. To avoid the problem, it is feasible to apply various dispersed devices and ejection aeration combination with other methods, for instance, aeration mechanical system. It allows not only to increase aerated volume but to achieve tiny bubbles.

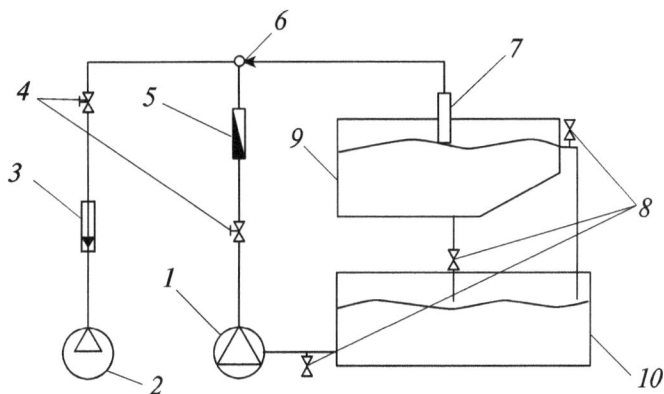

Fig. 5.21. Scheme of laboratory device:
1 — pump; 2 — compressor; 3 — rotamer; 4 — regulating valve;
5 — water expenditure counter; 6 — ejector; 7 — aerator;
8 — globe tap; 9 — flotochamber; 10 — initial water tank

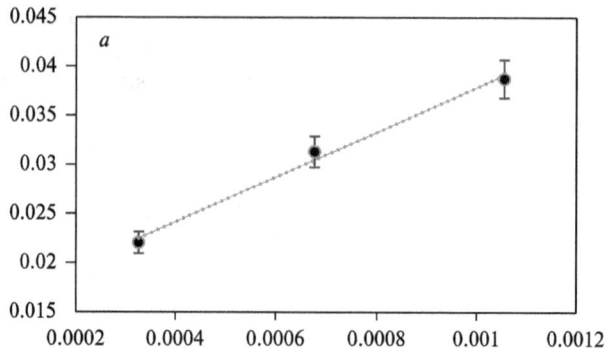

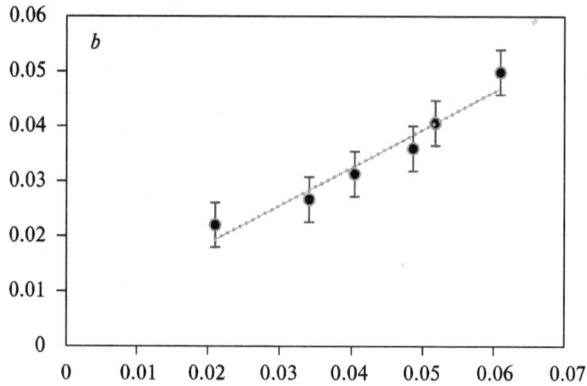

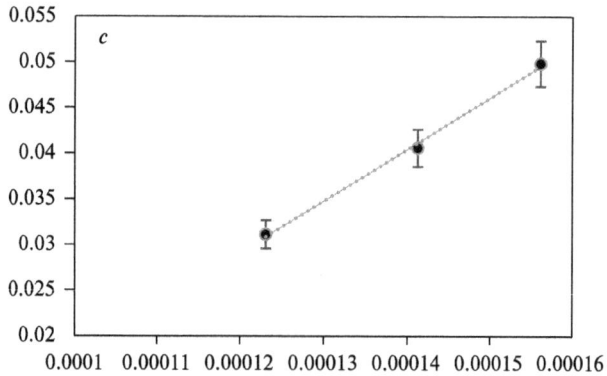

Fig. 5.22 Aeration intensity dependence from air expenditure:
a — water expenditure — 0.14 1/s; *b* — water expenditure —
0.18 1/s; *c* — water expenditure — 0.22 1/s

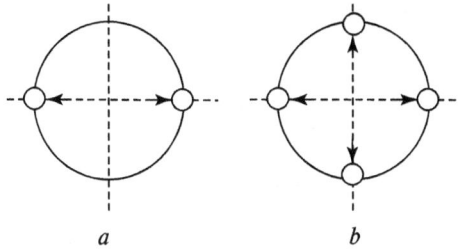

Fig. 5.23. Ejector location possible schemes:
a — 2 ejectors; b — 4 ejectors

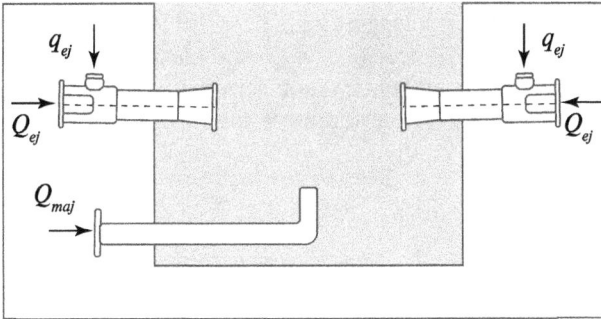

Fig. 5.24. Calculated scheme of flotation skimmer
with aeration ejection system

Let give an example of calculation.

The expenditure of wastewater given for clearing — Q, m³/h.

Experimental data testify that air necessary quantity for the effective purification is till about 15% from wastewater supplied end volume. Thus, air necessary quantity is calculated by the formula:

$$q = 0.15Q, \text{ m}^3/\text{h}.$$

Calculated scheme is presented on Fig. 5.24.

Necessary quantity of the air, supplied to one ejector, is estimated by the formula:

$$q_{ej} = \frac{q}{n}, \text{ m}^3/\text{h},$$

where n — ejector number in apparatus (usually 1, 2, 4 or more depending on wastewater volume).

Then, it is calculated the quantity of wastewaters, supplied to one ejector, for to provide inleakage of air necessary number:

$$Q_{ej} = \frac{q_{ej}}{k}, \text{ m}^3/\text{h},$$

where k — coefficient of ejecting capability of ejector.

Let determine afterwards wastewater expenditure through major supply sleeve:

$$Q_{maj} = Q - Q_{ej} \cdot n, \text{ m}^3/\text{h}.$$

The attention should be paid on that recirculation of purified wastewaters to avoid ejectors' plugging can be provided in flotation machine with ejection aeration. Hence, $Q_{maj} = Q$, and cleaned wastewater quantity, supplied repeatedly into apparatus via one ejector, — Q_{ej}.

Then we determine working volume of flotation skimmer:

$$V_{work} = Q_{work} \cdot t, \text{ m}^3,$$

where $Q_{work} = Q_{maj} + Q_{ej}$, m^3/h.

Calculation example: the rate of wastewaters, supplied into apparatus for cleaning, is 5 m^3/h. Let accept ejector number in apparatus as 2 with ejection coefficient as 0.5.

Air necessary quantity for purification:

$$q = 0.15Q = 0.15 \cdot 5 = 0.75 \text{ m}^3/\text{h}.$$

For one ejector:

$$q_{ej} = \frac{q}{n} = \frac{0.75}{2} = 0.375 \text{ m}^3/\text{h}.$$

The quantity of wastewaters, supplied to one ejector, for inleakage of air necessary volume:

$$Q_{ej} = \frac{q_{ej}}{k} = \frac{0.375}{0.5} = 0.75 \text{ m}^3/\text{h}.$$

Wastewaters expenditure via major supply sleeve:

$$Q_{maj} = Q - Q_{ej} \cdot n = 5 - 0.75 \cdot 2 = 3.5 \text{ m}^3/h$$

Let accept water time-being in apparatus as 0.5 h. Then working volume of flotation skimmer:

$$V_{work} = Q_{work} \cdot t = 5 \cdot 0.5 = 2.$$

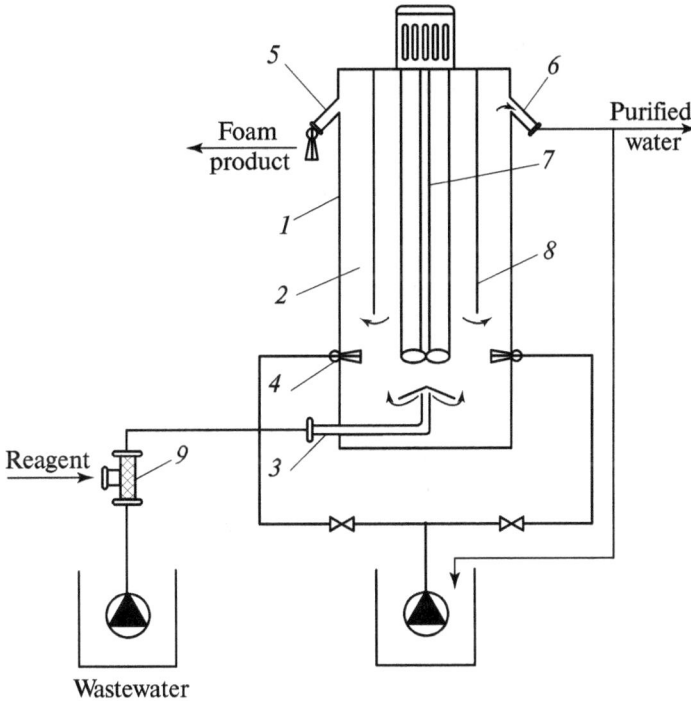

Fig. 5.25. Scheme of flotomachine with ejection aeration ejection
system and impeller

Installation of aeration combined system, presented on Fig. 5.25, is
developed as an example of ejection aeration in flotation purification.

Installation includes Shell *1*, which external side has Sleeve *3* of dirty water supply, Sleeves *4* with installed ejectors for water-air-mixture supply as well
as Sleeves of bend of foam product *5*, purified water *6* whereat there are the
contours of wastewater supply and recirculation.

Flotomachine work principle comprises the following. Original dirty water, mixed with reagent in Mixer *9*, via Sleeve *3* comes into Flotochamber *2* of
Shell *1* of the device where also purified water, saturated with air, is supplied
via Sleeve *4*. Forming mixture of water and reagent is being mixed with the
help of impeller Block *7* which also provides additional crushing of bubbles
and their even distribution along flotochamber. The given mixture comes into
space, limited by Envelop *8* where the lift of formed flotocomplexes proceeds.
Flotomud in the form of foam layer is removed via Sleeve *5*, and purified water — via Sleeve *6*. Then, purified water half rate is directed on ejectors along
the contour.

One of the most important nodes of given installation is aeration system presented by impeller Block 7 and Ejectors 4.

In the installation, being considered, purified water is supplied on ejectors for to avoid their plugging. Water-air mixture input is held along apparatus axis.

Attention should be paid on that given device provides for reagents input. Reagent comes into Mixer 9 which in fact presents itself a tube with incorporated swirler of flow.

Depending on wastewater character, the two systems of original mixture distribution are provided by apparatus: for wastewaters with high content of fatty pollutions of big size, the water distribution is made with the help of bonnet. If pollution concentration is insignificant, the possibility to use disk aerator is envisaged that simplifies given node design.

Wastewaters, mixed with reagents, then come onto mixer which provides for additional blending of reagent with water as well as with air bubbles. Thus, possibility of complex (pollution particle)-bubble-reagent formation rises. Moreover, such device not only provides for reagent usage but also allows to exclude the node of reagent solution mixing.

5.6 EJECTOR AS A BLENDER OF A REAGENT WITH WASTEWATER

Reagent treatment of water is one of the most spread ways of water purification intensification. Important significance in reagent treatment of water is not only in the selection of reagent dose but also in the process of mixing of reagents, in particular, coagulants, with water. Coagulation process proceeds rather quickly and, in this regard, it is important to distribute reagent in water most evenly and quickly. The intensity of reagents mixing with water is characterized by speed gradient value G (c^{-1}) as well as by Camp criterion.

$$Ca = G \cdot \tau,$$

where τ — mixing time, s.

Speed gradient can be found by formula:

$$G = \sqrt{\frac{W}{\mu}},$$

where W — wattage, spent for water mixing, referred to volume unit in mixing chamber;
μ — dynamical viscosity.

Reagent

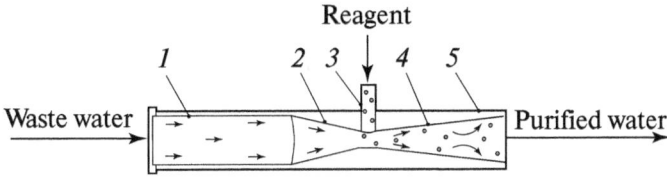

Fig. 5.26. Ejector scheme:
1 — entry chamber; 2 — nozzle; 3 — inleaked sleeve;
4 — mixing chamber; 5 — ejector corps

Blending time is determined by the formula:

$$T = \frac{V}{Q},$$

where Q — expenditure of wastewaters, m^3/s;
V — blender (mixer) working volume, m^3.

In the case of preliminary pursued optimization of mixing process, the following condition should be fulfilled:

$$T = r.$$

The accomplishment of aforementioned relations gives possibility to hold the process of reagents mixing with water practically in optimal regime.

Quick mixing of reagents with water, as a rule, raises the effectiveness of their usage. Nevertheless, the application of apparatuses with mixers for these purposes as the most effective technical means leads in the series of cases to the destruction of being formed aggregates whereat, that demands in this regard the strict following a temporarily interval of mixing. The simplest in apparatus regard and the effective device at the same is an ejector [33, 34]. In the being considered case, ejector, functioning as a jet pump, allows to dose and blend reagent with wastewater (Fig. 5.26).

For to investigate the process of reagent blending with wastewater in ejector chamber, the ejector testing model, presented on Fig. 5.27 has been chosen.

Model has been projected and researched in program complex ANSYS (further as Program) with the statement of model work various parameters. Experiments' series was based on single-phase model usage, i.e. the program was making calculation with account of similar liquid supply into both sleeves. Heat exchange functions (energy) haven't been considered in the experiment frames. Turbulence has been taken into account.

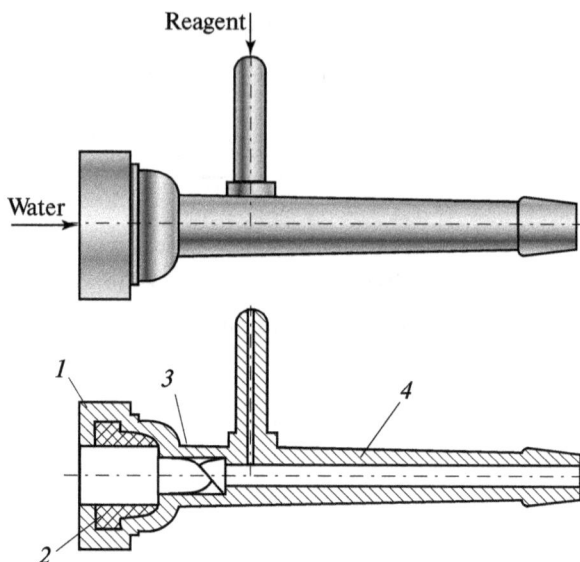

Fig. 5.27. Ejector design specificities:
1 — ejector cover (plastic); *2* — rubber strip; *3* — curved metallic
wall for twisting of *1* mm depth flow; *4* — ejector corps (plastic)

The calculation has shown quick and stable exit into work stationary regime, the time constituted 60 sec.

Major parameter of ejector application efficiency investigation as a mixer for reagent solution supply into being purified wastewater flow is indicators of speed (and its variations).

Fig. 5.28–5.33 show iso-surfaces of liquid flow speeds which give visual demonstration on what indicators of speeds prevail on individual sites inside ejector chamber.

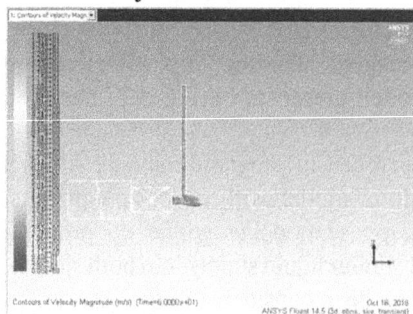

Fig. 5.28. Iso-surface $V = 12$ m/s

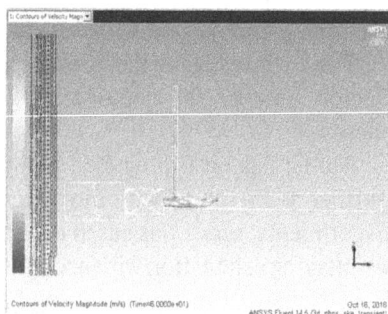

Fig. 5.29. Iso-surface $V = 10$ m/s

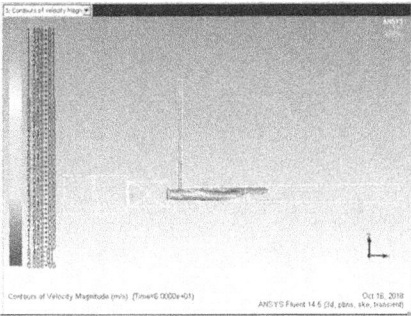

Fig. 5.30. Iso-surface $V = 8$ m/s

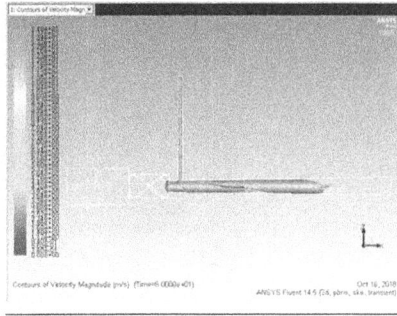

Fig. 5.31. Iso-surface $V = 6$ m/s

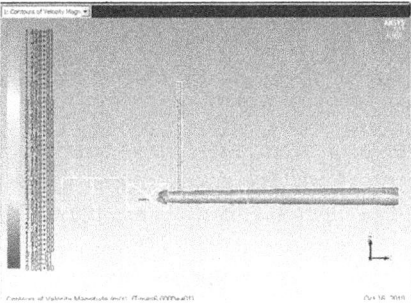

Fig. 5.32. Iso-surface $V = 4$ m/s

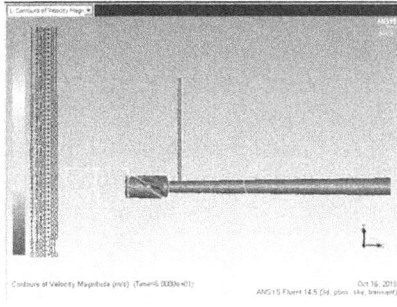

Fig. 5.33. Iso-surface $V = 2$ m/s

Distribution contours of speeds (speed magnitudes) on all sites of ejector mixing chamber in corresponding planes are shown on Fig. 5.34–5.35.

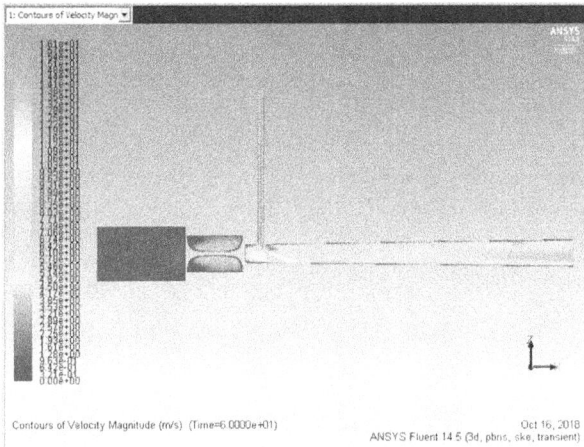

Fig. 5.34. Speed magnitudes in *XOZ* section

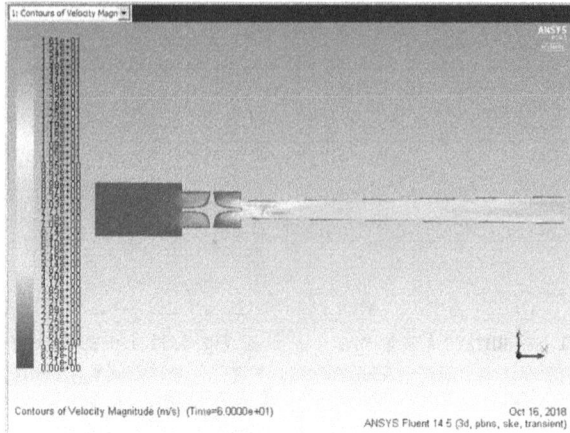

Fig. 5.35. Speed magnitudes in *XOY* section

As it seen from figures, the most intensive change in speed goes on the site of the input of liquid (reagent) second flow. The graphs of distribution for speeds on given site of 3 cm length of blending chamber are presented on Fig. 5.39.

Abscissa axis represents itself the position of ejector section point in *ZOY* plane (perpendicular to section, presented on figures) from mixing chamber start. Whereat, position on axis of movement of mixed liquids major flow in ejector (ejector axis in *OX* direction) corresponds to coordinate 0.000.

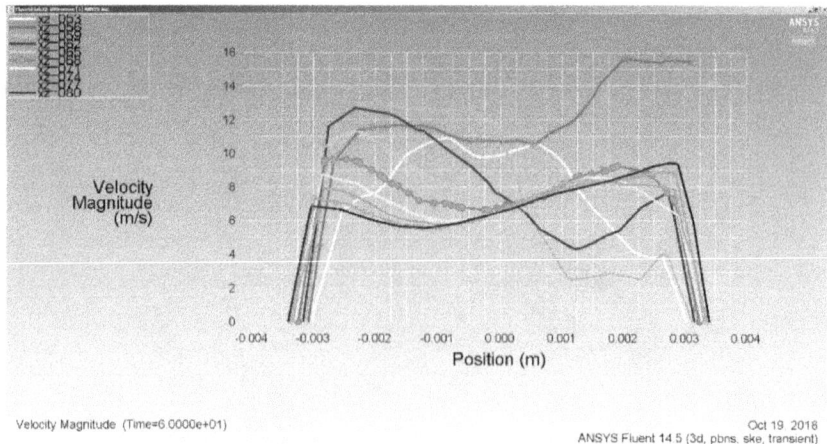

Fig. 5.36. Speeds distributions along mixing chamber

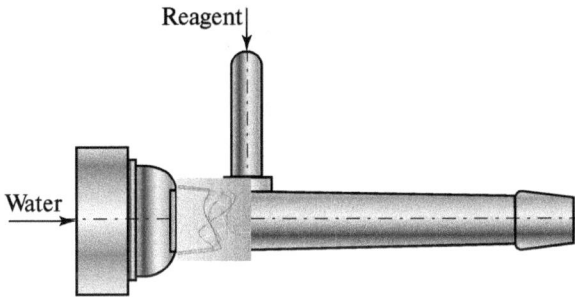

Fig. 5.37. Speeds distributions along mixing chamber

Speed values in m/s are shown along ordinate axis.

Meanwhile, the graph, corresponding to legend *XZ*_053, is white, characterizes speed magnitude distribution on the segment, remote from model left edge on 53 mm, that relates to the beginning of chamber of liquid immediate blending. The magnitude is given in *XOZ* section along ejector length.

Then magnitude picturing and graph building is made with 3 mm step from mixing chamber beginning.

It is seen on the figure that distribution speed graph *XZ*_053 (red) doesn't allow "sticking" to ejector wall, i.e. it is the place of reagent input (ejector sleeve).

The position of this graph of speeds inside ejector corps is more demonstratively shown on Fig. 5.37.

But single-phase model doesn't give full presentation on processes which will go on in blending chamber while the introduction into being purified wastewater of reagent solution via ejector sleeve, intended for air suction (in the case of ejector classical application), as being mixed liquids will have differing characteristics from each other and, respectively, they should be presented as 2 different flows for calculation in ANSYS. Whereat, there is no necessity to set reagent individual composition and precise characteristics of being applied liquid on the research given stage, it's sufficient to set two-phase model of calculation in the program.

Two-phase calculation model was set in experiments' second series. Meanwhile, media (liquid) homogeneity has become one of the researched parameters. Homogeneity, in given case, is directly connected with wastewater purification degree as the better reagent (flocculant, coagulant) is spread in being cleaned wastewater the higher flotation purification efficiency will be.

Liquids volume distribution (phase mixing) is shown on Fig. 5.38–5.39.

121

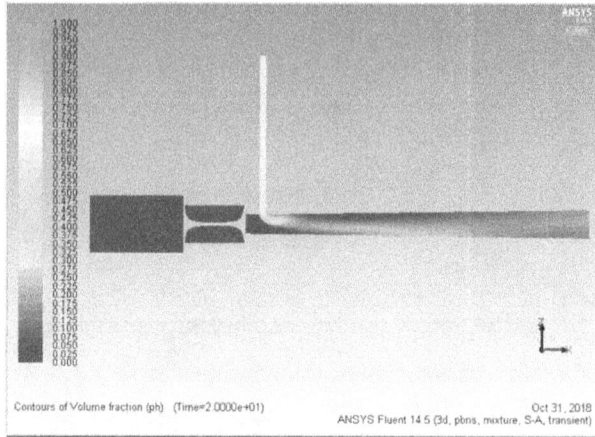

Fig. 5.38. Volumetric distribution of phases (mixing), *XOZ*

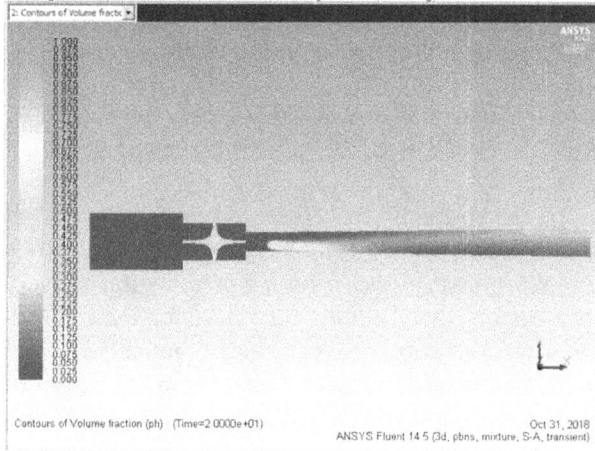

Fig. 5.39. Volumetric distribution of phases (mixing), *XOY*

It is seen from the distribution that for mixing process intensification, it is needed to twirl flows more, for instance, by the construction change of metallic wall on main flow entrance into ejector, or to increase the speed of liquid flow supply in reagent input sleeve. Another alternative intensification method of the process of blending and the obtaining on exit from ejector of more homogeneous liquid is the increase of length of ejector itself.

One of the major parameters, characterizing mixing process, is speed ingredient. Table 5.8 shows speed gradient dependence from water supply initial speed.

Table 5.8. Speed gradient dependence from water supply initial speed

Water speed on entrance, m/s	Expenditure, kg/s			Speed of, m/s			Nexit, %	Speed component gradients, s^{-1}		
	Water	Reagent	Residual	reagent on entrance	mixture max.	mixture on exit		V_x max	V_y max	V_z max
0.4 (7s)	0.102	0.003	−4.7e−06	0.919	5.829	1.645	2.76	12776	6188	7065
0.8 (14s)	0.203	0.007	−1.0e−05	2.115	11.634	3.304	3.16	24162	11931	14019
1.2 (21s)	0.305	0.010	−1.5e−05	3.310	17.452	4.962	3.29	36226	17920	20988
1.6 (28s)	0.406	0.014	−2.0e−05	4.487	23.273	6.620	3.35	48302	23906	27986
2.0 (35s)	0.508	0.018	−2.4e−05	5.655	29.094	8.277	3.37	60406	29897	34999
2.4 (42s)	0.609	0.021	−2.8e−05	6.811	34.920	9.934	3.39	72532	35896	42029

The analysis of data, presented in Table 5.8, shows that speed gradient along X axis and Y and Z axes essentially depends on water supply initial speed. Whereat, the ratio of speed gradient along X axis to values on Y and Z axes constitutes approximately 2:1. It demonstrates that media turbulization doesn't differ drastically by pointed directions that, in our opinion, must provide for rather good mixing of reagents with purifying water. Not only reagent dose selection but also reagent mixing process, in particular, of coagulants, with water has important significance at water treatment with reagents. Coagulation process proceeds rather quickly and, in this regard, it is important to distribute reagent in water more evenly and quickly.

Reagents quick mixing with water, as a rule, increases efficiency of their use. Nevertheless, the application of apparatuses with mixers for these purposes, as the most effective technical means, leads in the row of cases to the destruction of whereat forming aggregates that demands, in this regard, strict following of mixing time interval. The simplest in apparatus regard and effective device at the same is an ejector. Ejector, functioning as a jet pump, allows to dose and mix regent with wastewater.

For to check proposed technical settlements, an experimental installation has been created.

Such installation principal scheme is presented on Fig. 5.40. Wastewater is supplied into Ejector *1* where, in narrow place, in underpressure points, reagent solution is pumped from Reservoir *2*. Mixing of solution with wastewater proceeds in mixer chamber, whereupon flow comes into Reservoir-Decanter *3*.

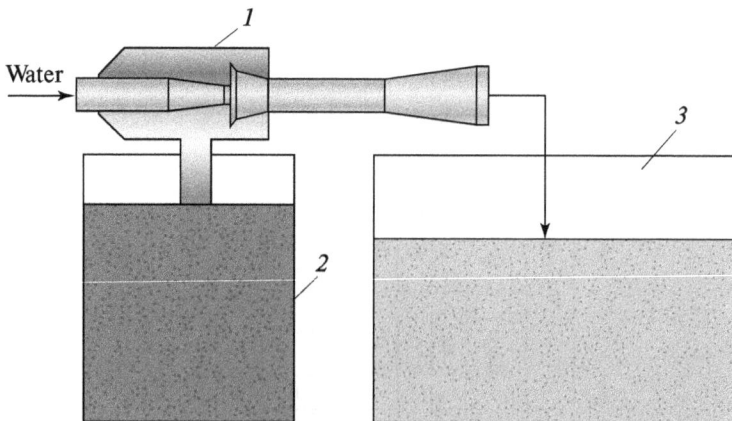

Fig. 5.40. Scheme of installation with ejector-dispenser:
1 — ejector; *2* — reservoir with reagent solution;
3 — reservoir-decanter

Experiments have shown that at water supply in the quantity of 1 l via ejector, 20–50 ml of reagent solution is being in-taken during 30 seconds (Fig. 5.41).

In course of research holding, empirical dependence was established:

$$q = k \cdot Q, \qquad (5.15)$$

where q — quantity of pumped solution of reagent;

Q — processed water quantity;

k — proportionality coefficient.

Coefficient k in equation (5.15) has rather wide interval of values and this value ruling is achieved by the way of change by technical approach of solution q supply, for example, by the way of regulation by valve or any other device.

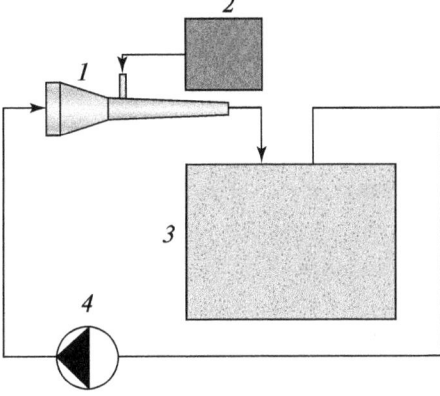

Fig. 5.41. Ratio between treated water and pumped reagent solution:
1 — quantity of water, supplied into ejector,
2 — quantity of pumped reagent solution

To pursue the research, more complex installation has been used. Such installation principal scheme with dosing device is pictured on Fig. 5.42. Wastewater with Pump 4 is supplied into Ejector 1; in wastewater, in underpressure zones, reagent solution is pumped from Reservoir 2.

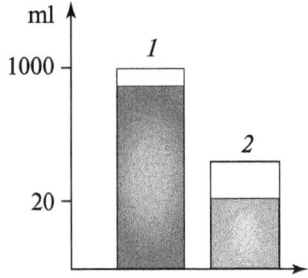

Fig. 5.42. Laboratory installation scheme:
1 — ejector; 2 — reservoir with reagent solution; 3 — decanter;
4 — pump

In ejector mixing chamber, solution blending with wastewater proceeds, whereupon the flow is transferred into Reservoir-Decanter 3. Ejector is installed on tripod with the possibility to regulate jet height and fall angle. Then water in the form of jet returns into 3, thus, water circulates continually.

Several kinds of flocculants of Nalco brand with 0.05–0.6 % of water volume were used in given investigation as well as coagulant Aqua-Aurate — 30–5% solution.

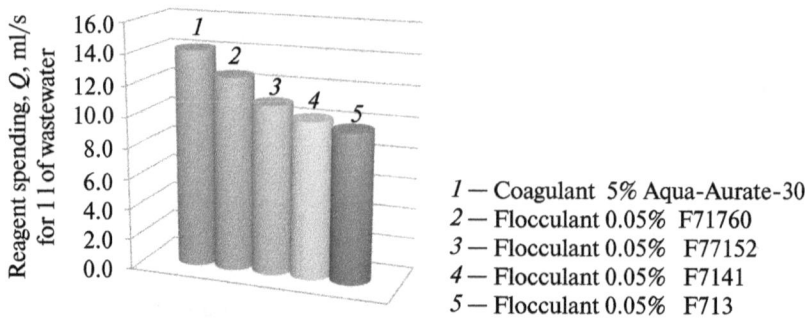

1 — Coagulant 5% Aqua-Aurate-30
2 — Flocculant 0.05% F71760
3 — Flocculant 0.05% F77152
4 — Flocculant 0.05% F7141
5 — Flocculant 0.05% F713

Fig. 5.43. Diagram n.1 (reagent quantity to wastewater of 1 l)

Table 5.9. Experimental data on definition of expenditure of reagent to 1 l water flow

	V, ml		t, s	Q, ml/s	
	reagent	wastewaters		reagent	wastewaters
Coagulant Aqua-Aurate	90	1000	6.35	14.2	157.5
Flocculant 0.05% 7141	65	1000	6.35	10.2	157.5
Flocculant 0.05% 71760	80	1000	6.35	12.6	157.5
Flocculant 0.05% 713	62	1000	6.35	9.8	157.5
Flocculant 0.05% 77152	70	1000	6.35	11.0	157.5

Rate of reagent, given into ejector, is determined under the formula:

$$Q = \frac{V}{t},$$

where V — volume of solution of sucking reagent, ml, t — time, s.

Experiments have shown that ejector with short mixing chamber has weak suction of reagent solution, therefore given type of ejector hasn't been used for the study.

At water supply via ejector with elongate mixer chamber of 1liter quantity, reagent solution 62–90 ml is being pumped during 6.4 seconds (Fig. 5.43 — diagram n.1). Experimental results are demonstrated in Table 5.9. Hence, these data are applicable in the case of industrial installations.

For flocculants of various viscosity, there have been held lab investigations of the change of pumped reagent rate depending on solution concentration. The results are presented on Fig. 5.44.

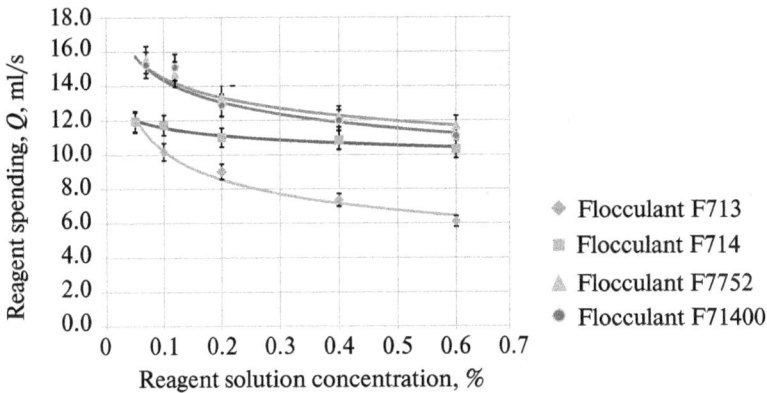

Fig. 5.44. Reagent rate dependence from concentration

The analysis of data, presented on Fig. 2.8–2.9, shows that the productivity on reagent supply directly depends on coagulant and flocculant concentration. Productivity of more viscous flocculant solutions falls. Therefore, necessity occurs to pursue correction of suction rate value in view of the density of pumped reagents and ejector height relatively pumped liquid level in service tank. Reagent suction productivity correction has been made by the formula:

$$q = \frac{q_{theor}}{\gamma},$$

where q — real rate of suction of reagents;

q_{theor} — suction theoretical rate in view of that pumped liquid is a water;

γ — reagent specific weight, g/dm^3.

There were also held the comparative research of ejector characteristics in dependence of device immersion into water. As diagram n.2 shows (Fig. 5.45), discrepancies in reagent rate values, depending on ejector immersion depth into water, vary slightly. Nevertheless, the use of freely falling jet allows to increase pumped reagent quantity significantly.

Research and calculations were held taking into account that ejector is installed on one and the same height with reagent suction level. In other cases, it is necessary to hold reagent rate correction, taking into account difference level, according to the formula:

$$q = \frac{q_{theor}}{1 + 0.1\left(H_r - H_{ej}\right)},$$

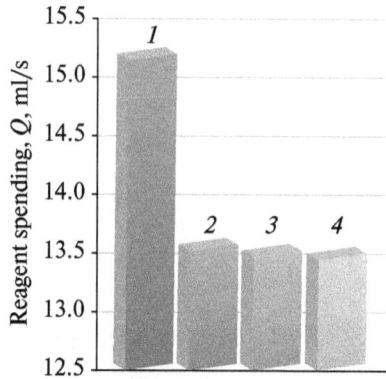

Fig. 5.45. Diagram n. 2 (reagent expenditure depending on ejector immersion depth into water):
1 — free stream; *2* — immersion depth 1 cm;
3 — immersion depth 3 cm; *4* — immersion depth 5 cm

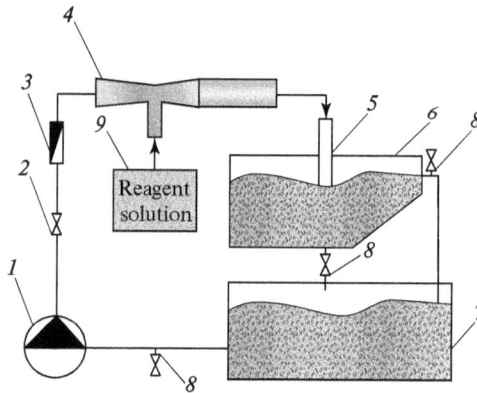

Fig. 5.46. Scheme of installation with jet aerator and ejector:
1 — pump; *2* — regulating valve; *3* — water rate counter;
4 — ejector; *5* — aerator; *6* — flotochamber;
7 — initial water tank; *8* — globe-faucet

where q — real rate of reagents suction;
q_{theor} — suction theoretical rate;
H_r — reagent level height, m;
H_{ej} — ejector installation height, m.

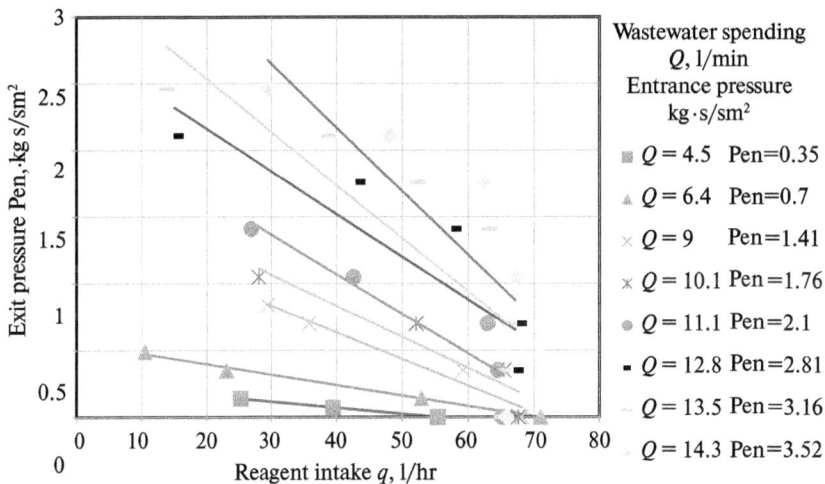

Fig. 5.47. Productivity dependence on reagents suction versus pressure on exit

Legend for Fig. 5.47:

Wastewater spending Q, l/min
Entrance pressure $kg \cdot s/sm^2$

- $Q = 4.5$ Pen=0.35
- $Q = 6.4$ Pen=0.7
- $Q = 9$ Pen=1.41
- $Q = 10.1$ Pen=1.76
- $Q = 11.1$ Pen=2.1
- $Q = 12.8$ Pen=2.81
- $Q = 13.5$ Pen=3.16
- $Q = 14.3$ Pen=3.52

Axes: Exit pressure Pen, $kg\, s/sm^2$ vs. Reagent intake q, l/hr

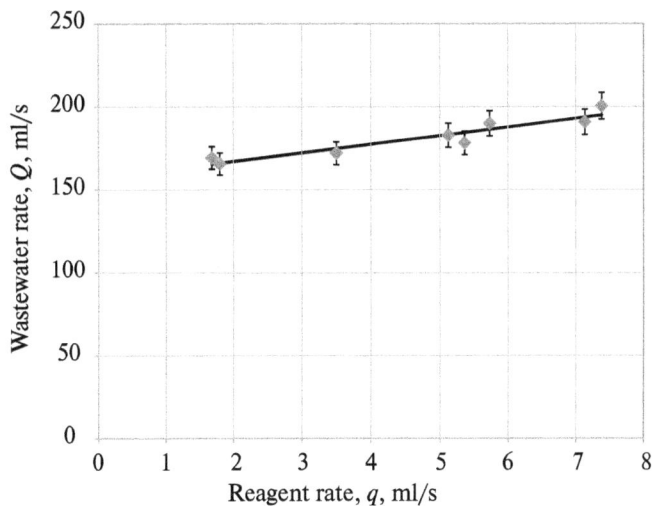

Fig. 5.48. Reagent suction dependence from wastewater rate

Axes: Wastewater rate, Q, ml/s vs. Reagent rate, q, ml/s

Experiments have been held on the device "Aerotank-decanter with jet aerator" for more detailed research of proportionality coefficient k from formula (5.15).

129

There have been also held investigations on the device comprising two major parts: aeration pool and jet aerator with ejector (Fig. 5.46).

Important results in our opinion have been got for practice (look Fig. 5.47 and 5.48) including for coagulant and flocculant dosing into wastewaters.

Thus, linear dependence of reagent solution pumping versus pressure on exit and water expenditure has been determined.

There is also interest in the consideration of the intensification process of reagent mixing with being cleaned water with the use of ejector, including magnetic field application.

Such ejector — is a blender worked out by us [30], includes shell, consisting of entrance chamber, which external side on are sleeves for working liquid, air supply and mixing chambers with exit sleeve, at that, entrance chamber is additionally furnished with sleeve for reagent solution supply; on mixing chamber externatl side, there are magnetic field sources and, whereat, entrance chamber length L0 and mixing chamber length L1 ratio is 1:3 till 3:50 and, whereupon, as sources of magnetic field, constant magnets and electro-magnets are used, they create magnet field intensity in blending chamber from 500 to 7000 Oersted with magnetic field intensity gradient in blending chamber from 100 to 500 Oersted/cm. Fig. 5.49 demonsrates ejector-mixer scheme (Russian Federation patent on useful model "Ejector-mixer" № 183320. Author Ksenofontov B.S.).

Offered ejector-mixer (Fig. 5.49) includes Shell *10*, comprising entry Chamber *2* which external side on are Sleeves for working liquid *1* and air *3* supply and which also has Nozzle *4* inside, mixing Chamber *5* with exit Sleeve *7*, at that, entry chamber is additionally equipped with Sleeve *9* for reagent supply, and external side of blending chamber has magnet field Sources *6* and *8* and whereat, entrance chamber length L0 and mixing chamber length L1 ratio is 1:3 till 3:50 and, whereupon, as sources of magnetic field, constant magnets and electro-magnets are used, creating magnet field intensity in blending chamber from 500 to 7000 Oersted with magnetic field intensity gradient in blending chamber from 100 to 500 Oersted/cm.

Working principle of ejector-mixer is in the following. Initial working liquid is supplied under excessive pressure into ejector-mixer corps via entry Sleeve *1* into mixing Chamber *2*. Then, while work liquid movement via Nozzle *4*, pressure decrease below atmospheric one proceeds in the liquid that leads to pumping via Sleeve *3* of air and of reagent solution, for instance, in the form of salts of aluminum, iron, via Sleeve *9*. Formed in blending Chamber *5*, the mixture of working liquid and solution of reagent and air intensively interfuses and then is being withdrawn via Sleeve *7*. Intensification of

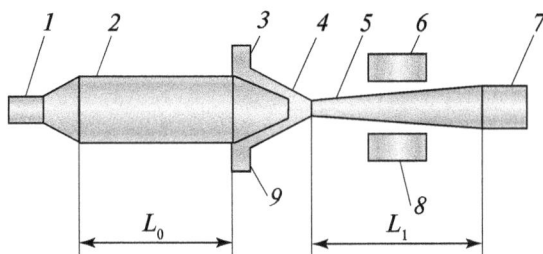

Fig. 5.49. Ejector-mixer scheme

agitation processes is promoted by definite ratio of entrance chamber length L0 and mixing chamber length L1 being from 1:3 till 3:50 and by the use as Sources of magnetic field *6* and *8* of constant magnets and electro-magnets, creating magnet field intensity in blending chamber from 500 to 7000 Oersted with magnetic field intensity gradient in blending chamber from 100 to 500 Oersted/cm. Meanwhile, ratio of lengths of entrance and mixing chambers contributes to turbulent diffusion gain, and magnetic field intensifies diffusion of ions of supplied reagents. Chosen magnetic field parameters are experimentally substantiated. In addition, lower limits are interfacial at the embodiment of ions diffusion strengthening effect, but above values of upper limits, the effect doesn't grow and the need on additional material and energetic costs for the maintenance of higher values of magnetic field intensity and intensity heterogeneity falls off.

At the accomplishment of declared conditions of ejector-mixer use in comparison of regular ejector (prototype), reagent usage efficiency rises on 20–30% leading to reagents' expenditure decrease, as well as material and energetic costs lower because of the absence of the need to apply pumps-dispensers in this case.

6 PRACTICAL APPLICATION OF FLOTATION MACHINES WITH VARIOUS SYSTEMS OF AEARATION

Taking in account of consequences, appearing at the usage of flotation multistage model, we have developed various technological schemes, approaches and flotation methods. Essential principled differences in flotation approaches are connected with saturation of liquid with air bubbles of definite size. Industrilal production most often uses machines and apparatuses of mechanical, pneumatic, presseure-head and electrical flotation. Last time, in Russia, abroad, multiple existing flotomachine designs have been improved that is why the largest interest reperesents the designs of renewed combined flotation devices. Let consider different types of flotation machines and apparatuses on the example of author's developments.

Ksenofontov B.S. has worked out mechanical flotation machines, in particular, FCP-0.15 (Flotation Combined Mechanical Purification Fig. 6.1) [11–12].

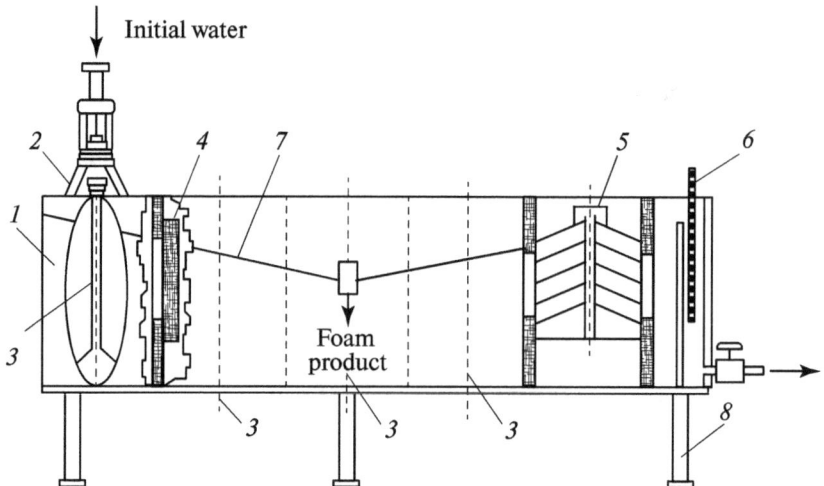

Fig. 6.1. Combined mechanical flotation machine:
1 — shell; 2 — aeration block; 3 — impellers; 4 — net;
5 — fin light; 6 — sunroof; 7 — foam trough;
8 — framework with stand

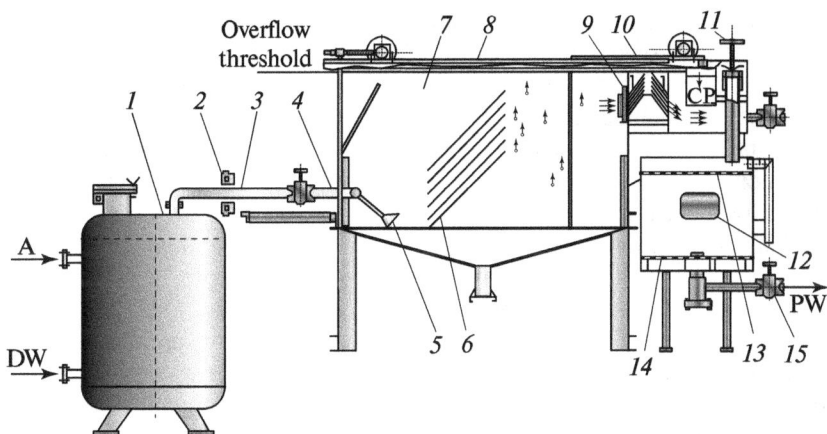

Fig. 6.2. Pressure-head flotation device:
1 — saturator; *2* — device for magnetic treatment; *3* — non-magnetic pipeline for water-air mixture supply; *4* — regulation device of water-air mixture supply; *5* — exit valve; *6* — block of inclined shelves; *7* — flotochamber workspace; *8* — device of foam takeoff; *9* — additional block of fine-zoned defecation; *10* — zone of output of flotocomplexes; *11* — water level regulation device in flotochamber; *12* — filter with granular load; *13* — distributor; *14* — drain system; *15* — valve; DW — dirty water; A — air; CP — caught pollutions; PW — purified water

This flotation machine differs from known ones by the presence of additional jet aerators in the form of collectors with tubes, which nozzles are installed inside, as well as of fin light in the form of the block of inclined shelves fixed at the distance of 5–10 cm one from another. Elements application rises aeration degree and decreases air bubble sizes, and fin light of special design allows to lower the gab of particles-microbubbles complexes that are late to lift to surface in previous chambers, thus, significantly increasing purification efficiency.

Similar principles for flotation machines and apparatuses development have been laid for pressure-head devices as well (Fig. 6.2).

Its difference from known analogs is in that air bubbles are being intensively extracted from the water; the air bubbles are located directly on impurity particles (drops) forming flotocomplexes which lift into foam layer. Complexes with small lifting power together with water flow get into 1st block of fine-zoned defecating where coalescence of air bubbles proceeds in natural conditions, and, hence, lifting force of these complexes proceeds. Then, being purified water gets into the second block passing preliminary through the net with hydrophobic surface, then through liquid level regulating device into filter with carbon load. Cleaning efficiency of such flotomachine can reach 95–98.

Improved electro-flotation apparatuses of combined type are used at the moment (Fig. 6.3) which distinctive feature is a mixing unit, fulfilled in the form of truncated pyramid.

Wastewaters come here together with reagent, preliminary introduced to treatment by induced electrical field on the tube site, occupied with electromagnet, then, there proceeds mixing and flotation of folliculated particles of fine-dispersed suspension on account of pseudo-liquefied layer from polymeric particles of 1.5–2.5 mm size between lattices. Then, flotopurification goes on in the chamber with impeller and fine purification — in electro-flotation chamber. Job efficiency of such apparatus regarding mineral oils can reach 99%.

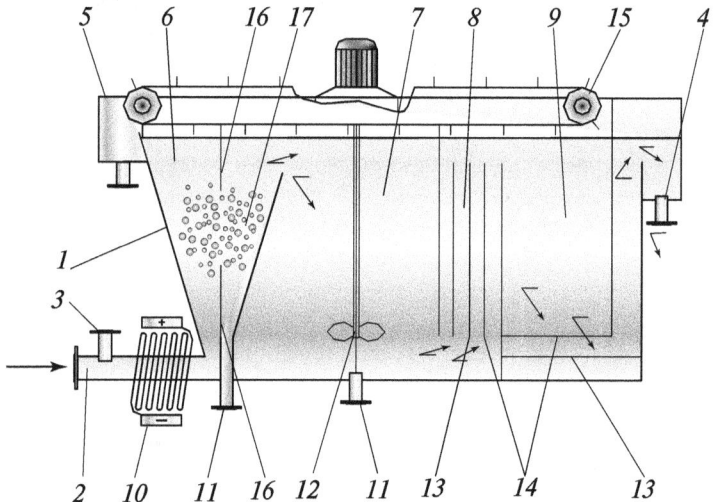

Fig. 6.3. Scheme of combined electro-flotation apparatus:
1 — shell of electro-flotation apparatus; 2 — entry sleeve for water supply; 3 — sleeve for reagent supply; 4 — exit sleeve for defecated water output; 5 — sleeve for foam product output; 6 — entry chamber; 7 — chamber of mechanical flotation; 8 — chamber with vertical electrodes; 9 — chamber with horizontal electrodes; 10 — electromagnet; 11 — sleeve for water supply; 12 — mixer; 13 — anodes; 14 — cathodes; 15 — foam rut; 16 — wall; 17 — pseudo-liquefied cap

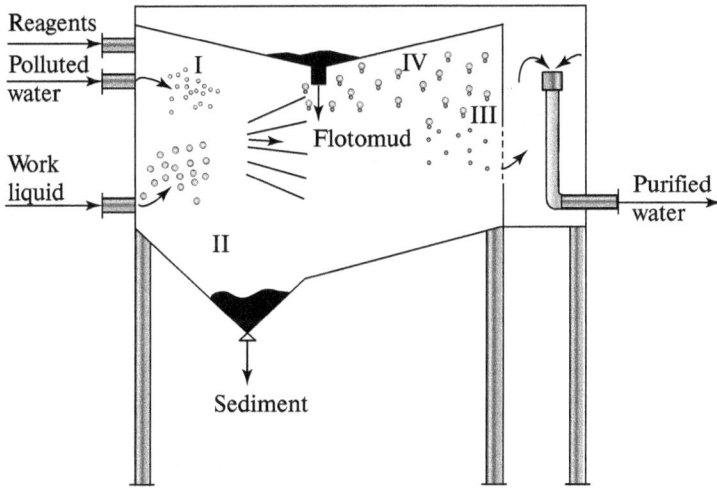

Fig. 6.4. Flotoharvester principled scheme

Whereat, it is worth to mention that flotation technique constantly improves as by the way of function combination as by the way of water purification degree rise. Conception, of developed by us flotoharvester for wastewater purification, corresponds to both directions of flotation technique improvement and, meanwhile, is a high-tech development that has been maintained by the big interest caused to this apparatus on various exhibiting events as well as the big interest caused at its practical usage.

Worked by us flotoharverster includes shell (Fig. 6.4) which external side on are sleeves for dirty water, reagents, working liquid (water with air bubbles) supply and sleeves for the bend of sediment, flotomud and cleaned water. Workspace inside the shell is divided on the zones of conditioning (I), defecating (II), coalescence of microflotocomplexes (III), immediate flotation (IV).

Flotoharvester working principle is in the following. Formation of flotocomplexes particle-bubble proceeds on condition stage (zone 1). Whereat, as a rule, not all pollution particles cling to air bubbles and, staying in sole state or in the form of aggregates, fall into sediment (zone 11).

Particles, stuck to small bubbles, form microflotocomplexes (zone 1) which slowly emerge on surface and, in this regard, are whisked by purified liquid flow, moving in horizontal direction. Such microflotocomplexes, reaching lattice wall, contact between each other with the formation of larger bubbles, as a rule, (zone 111) which quickly emerge on surface, forming flotomud (flotation zone IV). Being purified water, passing through lattice wall, is bent with the help special device and then withdrawn from flotoharvester via output sleeve.

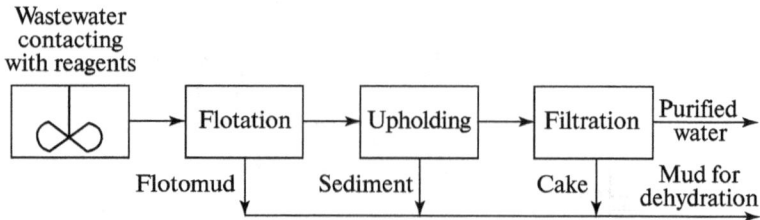

Fig. 6.5. Functional scheme of water purification in flotoharvester

Water cleaning effect in this case significantly exceeds reached results on devices-analogs.

Processes, taking place in flotoharvester, can be looked schematically with the help of the functional scheme presented on Fig. 6.5. Additionally, in defecating chamber, there can be installed mixing device for dirty water contacting with reagent solution and further there can proceed processes of flotation, defecating, filtering and other accompanying operations conducing to purification efficiency rise and to thickening of formed mud after purification of wastewaters (Fig. 6.5).

Such type flotoharvester pursued trials have shown its efficiency at the usage in wastewater purification processes.

At the necessity, process intensification can be achieved by the way of use of additional nodes, for example, by the way of installation of special blocks and so on. In individual cases, purification efficiency from the usage of pointed additional blocks can be raised approximately till 40–50.

To raise wastewater purification efficiency, we have developed different variants of flotoharvesters [50].

This workout technical result is the rise of wastewater purification efficiency and the obtaining of flotomud in thickened form.

As a result of use of new additional blocks in flotoharvester the quick bend of caught pollutions from internal space and their consequent thickening, leading eventually to purification efficiency raise at the expense of drastic decrease of particles fall from foam layer and to thickened sediment obtaining in one apparatus — flotoharvester, improve.

Let's show as an example the scheme of surface drainage with the use of flotodecanters being flotoharvester the simplest variant.

It's worth to note that from 2019 in RF it has been planned the usage of the best available technologies (BAT) for water flows purification, including surface drainage.

Last time, the cases of the falls of extremely strong rains have become more often that in some cases, leads to the flooding of residential territories, population evacuation, big material damage.

136

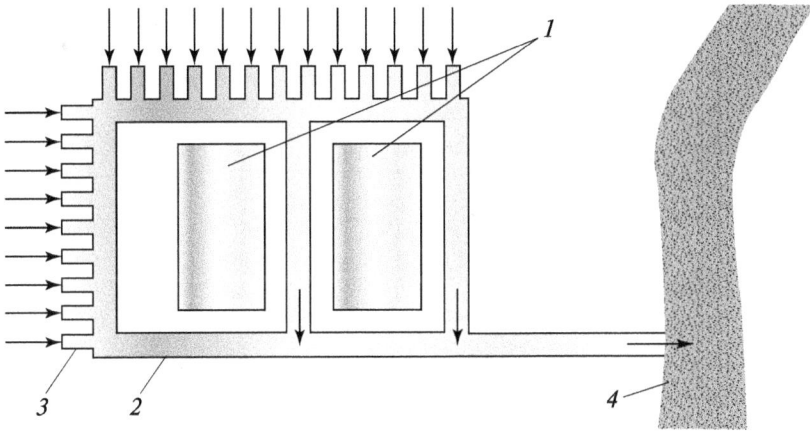

Fig. 6.6. Scheme of drainage from residential territory:
1 — production facilities; *2* — drainage channels;
3 — receiving canaliculi; *4* — river

Similar news from official media come almost weekly. Main reason for flooding is an absence or weak capacity of systems surface drainage.

Let consider possible cases of surface drain abstraction both from industrial or residential and undeveloped territories in regimes of heavy drainage. Figures 6.6–6.8 show the simplest examples of various variants of surface drain with the use of special drainage channels.

Example of the use of drainage ducts, which surface water comes to via collecting channels, is presented on Fig. 6.6. Heavy variant of surface drain abstraction is demonstrated on Fig. 6.7. In this case, the water comes into collecting ducts via all of their surface that strengthens water inflow into channels.

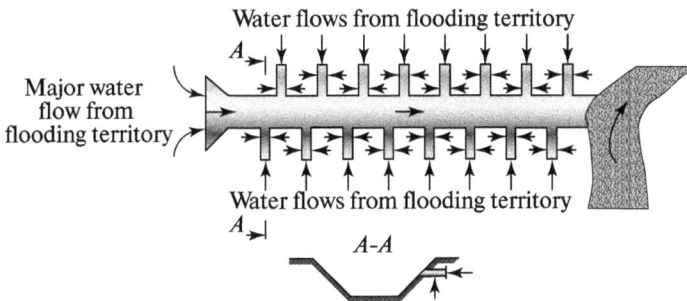

Fig. 6.7. Scheme of abstraction of surface drain from undeveloped territory during spring high water

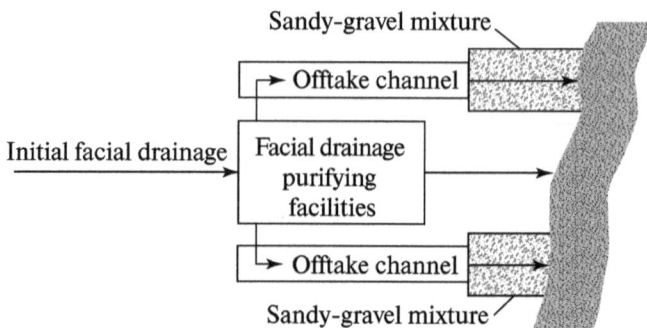

Fig. 6.8. Scheme of abstraction of surface drain from industrial territory during heavy rain fall or spring high water

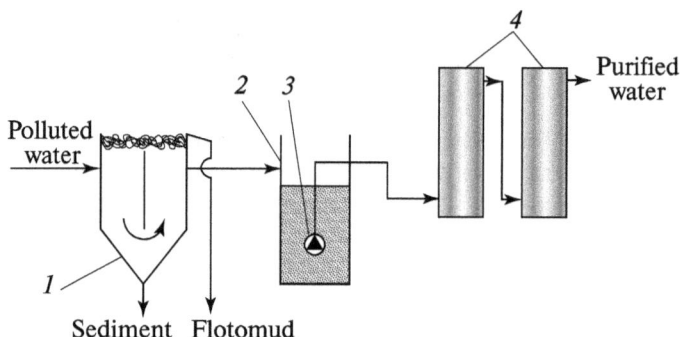

Fig. 6.9. Principled scheme of surface drain cleaning:
1 — flotodecanter; intermediate reservoir; *3* — pump;
4 — fine purification filters

Variant of heavy drainage and surface drain cleaning is presented on Fig. 6.8. In this case, the water is not only quickly delivered into abstraction system but is cleaned in special purifying facilities.

Facial drain abstraction in a whole should be planned strictly. In this regard, there are especially vital the problems of surface drain purification from residential territories see against intensive drainage. Effective technology of facial drain purification is developed by us with approximate scheme shown on Fig. 6.9.

It's known that facial wastewaters contain hydrophobic and hydrophilic pollutions in the majority of cases. In this case, it's feasible to use flotodecanters (Fig. 6.10, 6.11). In event of limited space for equipment installation it is possible to use column type flotodecanter (Fig. 6.10).

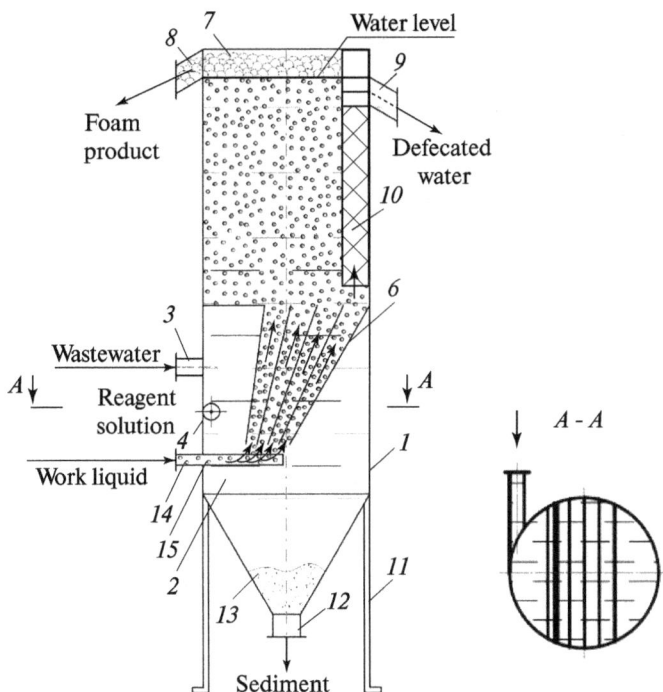

Fig 6.10. Scheme of column type flotodecanter

On the chance if there is enough space for equipment installation it's preferably to use flotodecanter with sequential chambers of flotation and defecating (Fig. 6.11).

Then preliminary defecated water comes into Flotochamber 7 where water cleaning proceeds from residual hydrophobic pollutions, for instance, mineral oils, oils and fats, on account of formation of flotocomplexes (dirt particle)-(air bubble) supplied in the form of mixture with water via tubular Aerator 14. Being formed floltocomplexes pollutions-(air bubbles) emerge on surface creating flotomud in the form of foam layer which is being deleted via Sleeve 8 and purified water is withdrawn via Filter 10 and then via Sleeve 9. Whereat, resettled pollutions are withdrawn via Sleeve 12 in the form of sediment.

Considered types of flotodecanters are rather essential factors of wastewater purification intensification. Defecating degree of wastewaters in flotodecanters can constitute up to 98—99 that is 10—15 higher than during the usage of flotation machines, for instance, of pneumatic type. Whereat, purification efficiency intensification can be achieved by the way of usage of reagent mixtures with exploitation of 3—5 kinds of reagents.

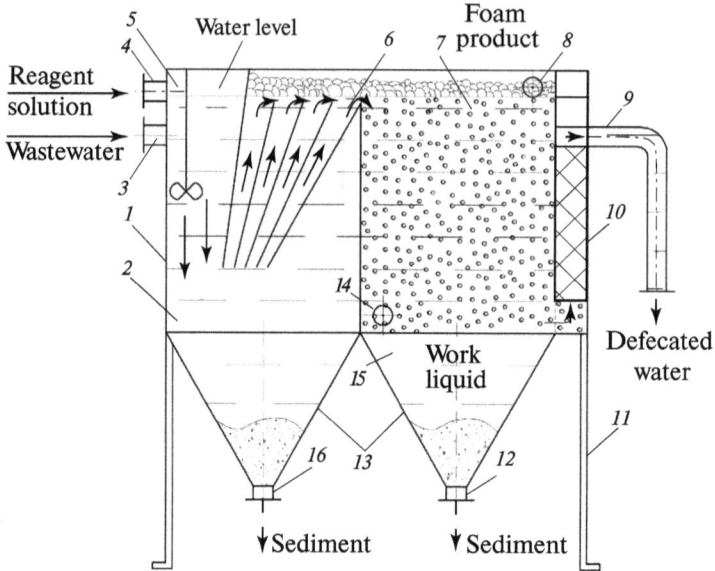

Fig 6.11. Scheme of flotodecanter with sequential chambers
of flotation and defecating

Flotodecanters work principle presented on Fig. 6.10, 6.11 is in the following. Initial wastewater via Sleeve *3* comes into Flotochamber *2* of Shell *1* of flotodecanter where reagent solution is supplied also via Sleeve *4*. Formed mixture of dirty water and reagent solution is mixed with the help of Mixer *5*. Blended mixture with occurring on the expense of reagent impact flakes from dirt particles enters into inter-shelves of Block *6* of fine-zoned defecating. Whereat, flakes, resettling at lower shelf, crawl down along inclined fin, striving into gutters of wavy material which these fins are made of. Then these flakes fall into sediment which is removed via Sleeve *16*.

After flotation cleaning, wastewater by drift comes into intermediate reservoirs wherefrom it is supplied by buried pump for fine purification stage, including filtering on mechanical and adsorption filters. In the process of the work of filters, in the volume of filtering load, the pollutions accumulate. For to remove caught compounds with the aim of service prolongation of filtering materials, the filters wash by water reverse-flow is provided.

As an adsorption filters load, the activated carbon of AG-3 brand is used. The activated carbon of AG-3 brand is a universal adsorbent of various polluting compounds from liquid and gas medias.

Carbon sorbs solute pollutions (detergents, mineral oils, solute organic compounds and others) on its surface in the job process. After the depletion of carbon adsorbed storage, it should be changed. The carbon load change is pursued with the help of hydro-unloading.

Reverse wash of carbon filters is provided in the system in the case of load change and for prophylactics. The wash is made with purified water, the washing waters are thrown into accumulating reservoir-decanter.

The water after sorption filters pours through ultraviolet sterilizer where disinfection of water proceeds, and it is collected in clean water reservoir wherefrom it is by drift dropped into city rain canalization.

Washing waters after filters regeneration are directed into the beginning of purification facilities — accumulating reservoir-decanter, whereupon they are exposed to repeating purification.

Results of the trials of given technology are demonstrated in Table 6.1.

Experimental data, shown in Table 6.1, demonstrate that as to major indicators (concentration of mineral oils and suspended substances), they satisfy the demands declared for the quality of purified wastewater being dropped into ponds.

Table 6.1. Results of the trials
of facial wastewater purification technology

Defined	Normative documentation for FIM (fulfilled instruction methodology)	Measurement results*, mg/dm^3		Measurement results violations**, mg/dm^3	
		Decanter	After filters	Decanter	After filters
Suspended compounds	NNDF (nature-protecting normative documentation of Federal level) 14.1:2.110-97	114.6	11.3	For all 10.0	For all 2.0
		120.1	11.1		
		112.4	11.2		
		116.6	10.8		
		117.8	10.5		
Mineral oils	NNDF 14:1:2:4.128-98	4.99	0.02	1.25	0.01
		1.48	0.02	0.37	0.01
		1.42	0.03	0.36	0.015
		1.49	0.04	0.37	0.02
		1.16	0.02	0.29	0.02

*Result has been obtained as average from two parallel definitions.

** Pointed discrepancies have been obtained in laboratory and supported by results of measurement accuracy control.

Whereat, specific energetic expenditures constitute 0.5–0.7 kW·h/m^3 and in the case of usage of known technologies (analogs) — till 1–2 kW h/m^3 and more.

Thus, the developed technology of surface wastewater purification allows to get cleaned wastewaters, meeting all demands, declared at their drop into opened pond, and can be considered as one of the variants of novel available technologies.

7 PROBLEMS OF SEDIMENT DEHYDRATION WITH THE USE OF FLOTATION HARVESTERS

The problem of processing of sediments, including excessive active silt, and their utilization during last decades is rather vital, and for now, universal ways for settlement of the tasks in this directions are not seen. Still multitonage wastes are utilzed in a small extent, and their major quantity is stored on polygons. In this regard, new technological settlements are extremely vital for practical technologies while processing and utilization of various wastes. Special significance in today's situation is in the settlement of engineering problems regarding dehydration of wastewater sediments, also including excessive active silt. Let consider additional functions of multifunctional flotoharvester (Fig. 4.2) meant as for wastewaters purification as for dehydration of wastewater sediments, including also excessive active silt. Let denote that earlier (Fig. 4.2) it was considered applying the use of apparatus for wastewater purification.

The developed flotoharvester [33] comprises shell, which external side on are sleeves for the supply of active silt original suspension, for the bend of defecated silt water and the withdrawal of flotomud and sediment, and perforated walls and defecated water withdrawal device are fixed up inside the shell. The flotoharvester whereat comprises additionally a sleeve for working liquid supply, a flotomud thickening block in the form of ejector, coarticulated with hydro-cyclone, and a sediment dehydration block, consisting of squeezing device which inside is the bag made of synthetic tissue, a tray with filter for filtrate collection is placed downwards. Whereupon, the ejector is fulfilled with the pressure drop in hereing in it from 0.05 to 0.5 m, and the synthetic tissue is accomplished with cell size from 0.001 to 0.1 mm.

The flotoharvester for active silt thickening (Fig. 7.1) includes Shell _1_, which external side on are Sleeves for active silt original suspension supply _2_, for silt defecated water bend _3_, foam Gutter _4_ with Sleeve _5_ of flotomud withdrawal, Ejector _6_ with Sleeve _19_ of compressed air supply and output Sleeve _9_, Hydro-cyclone _8_, Sleeves for thickened flotomud _10_, Sleeve _7_ of defecated liquid with bend Pipeline _11_, with Sleeves of sediment output _12_ and working liquid supply _21_, with additionally established Block _13_ of sediment dehydration on external side, the block consists of internal Chamber _14_ and external managing Devices _15_ and squeezing Clamp _16_. Meanwhile, Tray _17_ with inbuilt filter (shown with dotted line on Fig. 7.1) for defecated liquid collection is

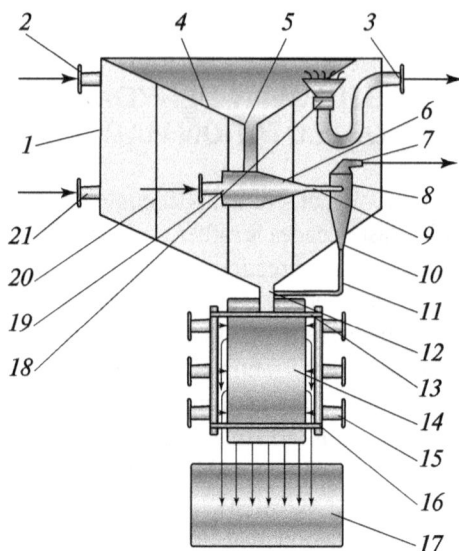

Fig. 7.1. Scheme of flotoharvester for division
and thickening of silt suspension
(RF patent on useful model № 72180; author Ksenofontov B.S.)

located in flotoharvester lower part, and perforated Walls *20* and purified water output Device *18* are installed inside flotoharvester.

As a result of new additional blocks usage in flotoharvester, quick bend of caught pollutions from internal space and their consequent thickening improves; it leads eventually to raise of silt suspension division on account of drastic reduction of fall-out of particles from foam layer and also leads to thickened sediment getting in one apparatus — flotoharvester.

Definition of optimal regimes of active silt treatment was pursued at different ratios of initial suspension of active silt and working liquid, as well as at variation of duration for treatment of mixture being divided. Experiment results are shown in Table 7.1.

Analysis of data presented in Table 7.1 shows that the best results have been achieved during active silt processing in flotoharvester at ratios of initial suspension of active silt and working liquid as 2:1 and duration of being divided mixture treatment during 20 minutes. At the use of such treatment regime, active silt biomass concentration is more than 6% of ADS (active dry substance) that makes such sediment transportable as well as in the case of necessity makes it applicable for the supply for thermal drying.

Table 7.1. Dependence of active silt suspension division efficiency from treatment regime in flotoharvester

Flotation regime		Biomass concentration, % ADS (absolutely dry substance))		
Ratio of initial suspension to working liquid	Duration of dividing, min	In initial suspension	In sediment	In defecated liquid
5.0:1.0	5	0.81	3.50	0.24
4.5:1.0	7.5	0.81	3.76	0.18
4.0:1.0	10	0.81	3.98	0,17
3.5:1.0	12.5	0.81	4.47	0.15
3.0:1.0	15	0.81	5.66	0.14
2.5:1.0	17.5	0.81	6.14	0.13
2.0:1.0	20	0.81	6.30	0.11
1.5:1.0	22.5	0.81	5.97	0.17
1.0:1.0	25	0.81	3.86	0.25
0.5:1.0	27.5	0.81	1.90	0.36

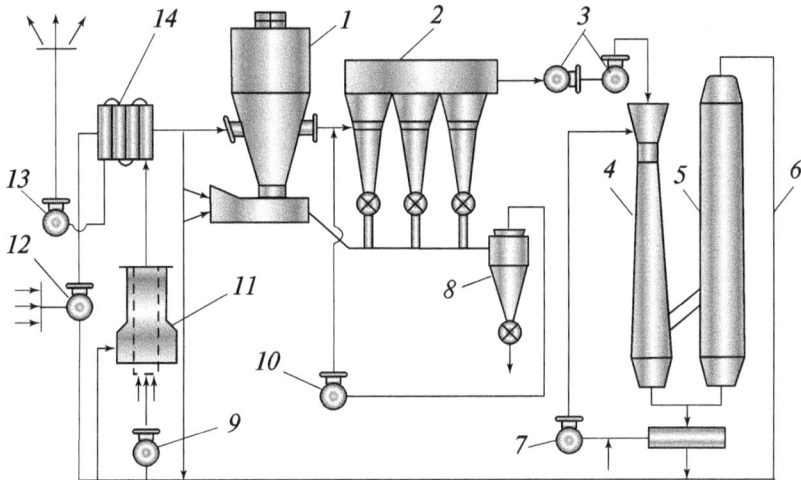

Fig. 7.2. Cleaning scheme of gas-air emissions of drying installation:
1 — drying chamber; 2 — battery of cyclones; 3 — exhauster;
4 — venturi pipe; 5 — dropper; 6 — flue; 7 — pump;
8 — cyclone; 9 — fan; 10 — blower; 11 — fire-box; 12 — fan;
13 — exhauster; 14 — air-heater

In the latter case, it's worth to use the drying with closed contour on thermo-carrier for the exclusion of toxic substances of mud into natural environment. Last matter is extremely important as many productions dry and burn muds without such detoxication.

Let consider technological scheme of the purification of gas-air emissions from drying devices (Fig. 7.2). Vapor-air mixture from battery of cyclones by two exhausters is given onto wet cleaning system, consisting of venturi pipes and two droppers, and goes into recirculation flue with condensate separators. Major part of vapor-air flow from the flue is directed by a fan together with atmospheric air into an air-heater and then as a thermo-carrier into a drying chamber. Left part of the flow is burnt in cyclone fire-box, further, it passes through air-heater and is emitted via smoke pipe of 120 m height. Thus, drying installation can work in closed cycle. Only that part of vapor-air flow is being emitted which has come through thermal treatment system.

Presented scheme of the drying guarantees the exclusion of getting of especially dangerous substances of muds into natural environment that is rather important, especially, during the drying of various liquid muds containing toxic compounds. Significance of presented technical settlements of thermal processing of wastes should be specially underlined because population, living closely to similar facilities, considers gas-purifying devices as low-efficient. In our case, the emission of polluting substances into atmosphere is excluded securely.

8 FLOTATION PROCESSES OF TARGET PRODUCTS EXTRACTION

8.1 POSSIBILITIES TO USE FLOTATION SKIMMERS FOR THICKENING AND INACTIVATION OF MICROORGANISM MASS

It's known that flotation approaches are used not only for wastewater purification and active silt thickening but also in biotechnological industrial processes, for instance, on hydrolysis plants, producing feeding yeast as on major production as well as for wastewater purification.

In this regard, being used flotation technique represents an interest. Flotation skimmers with aeration pneumatic system are used on hydrolysis enterprises for yeast biomass extraction. In the majority of cases, flotation skimmer has cylindrical corps which inside a glass for foam product pick-up is fixed up, and flotation process of yeast cells proceeds in a ring space between the corps and the glass, the space is divided into 4–5 sections. Whereat, there is a continuous wall between the first and last sections, not allowing for entry and exit flows to blend. Work principle of such apparatus is the following. Yeast suspension from yeast-plant apparatus comes into the first section which yeast cell flotation in proceeds at the expense of air, being in initial suspension. During yeast suspension coming through other sections, the flotation goes on account of the air, supplied via bubblers from a blower. Thus, continuous process of yeast cells extraction from liquid into foam proceeds at consequent pass of flotation skimmer sections. To slack foam layer, being created during foam layer flotation in flotation skimmer central part, a mechanical foam-extinguisher is installed upwards under the glass, for instance, in the form of a disk with a drive from an electro-engine. Being destructed by mechanical way, the foam comes into the glass and further — for a separating.

The use of flotation skimmer of such design for yeast suspension preliminary thickening is rather efficient and allows noticeably to decrease the quantity of separators used at further stage of thickening. Nevertheless, this flotation skimmer usage for suspension thickening of excessive active silt is not efficient and is practically inapplicable. For to use the flotation skimmer of above-described design for active silt thickening, as experimental research results show, is feasibly to reequip into a pressure-head flotation skimmer or to substitute aeration pneumatic system on a pneumo-hydraulic one. For that, first of all, it is necessary to reequip the flotation skimmer with an air saturation system, for instance, with a saturator or to embody air supply under pressure of 0.5–1.0 MPa into a pipeline for supply of active silt initial suspension.

Technological schemes of saturation of active silt suspension with saturator and with no saturator (in pipeline) have their own definite ups and downs. The advantage of the scheme with saturator is in that an air excess is separated out in saturator and not in flotation skimmer as in the case of the scheme without the saturator. Besides, more air in saturator is solved in liquid because an excessive pressure and time-being of silt suspension in saturator allow, as a rule, to embody this process in optimal regime. Drawbacks of scheme with saturator concern a particular strictness in demands for its accomplishment, whereupon following special technologies for the apparatus working at pressure, as well as a control for acceptance and further exploitation of saturator in the scheme of flotation thickening of active silt.

All of this, at definite measure, complicates both flotation device accomplishment in a whole and its exploitation.

Work scheme of flotation skimmer without saturator includes air supply under pressure into pipeline in front of a pump (Fig. 7.1, *b*). In this case, the necessity in making the saturator falls away that is an undoubtful advantage of the scheme. However, the excessive air, pumped with the help of ejector or directly given from compressor, is removed from suspension in the flotation skimmer, as a rule, in the form of large bubbles that worsens the process of flotation of active silt flakes. In this way, it is advisably to use special pumps for active silt supply or to create conditions, which air dissolving in silt suspension will go on with high speed at. It'll allow to increase spec. quantity of dissolved air in liquid phase and to avoid negative impact from air excessive emission in the form of big bubbles.

Analysis of afore-described approaches of active silt suspension saturation with air and results of experimental research testify that none of pictured scheme gives essential advantage. However, in the way of presence of full kit of flotation equipment, the scheme with saturator should be given the preference. Hereat, the matter of insoluble air removal is being settled. Besides, the possibility of water supply directly into saturator is given that noticeably prevents possible breaks in air supply with the help of ejector in the case of its chockage. While air supply into saturator, pump work proceeds without breaks. In the way of air ejecting into active silt suspension before the pump, work efficiency of the latter lowers.

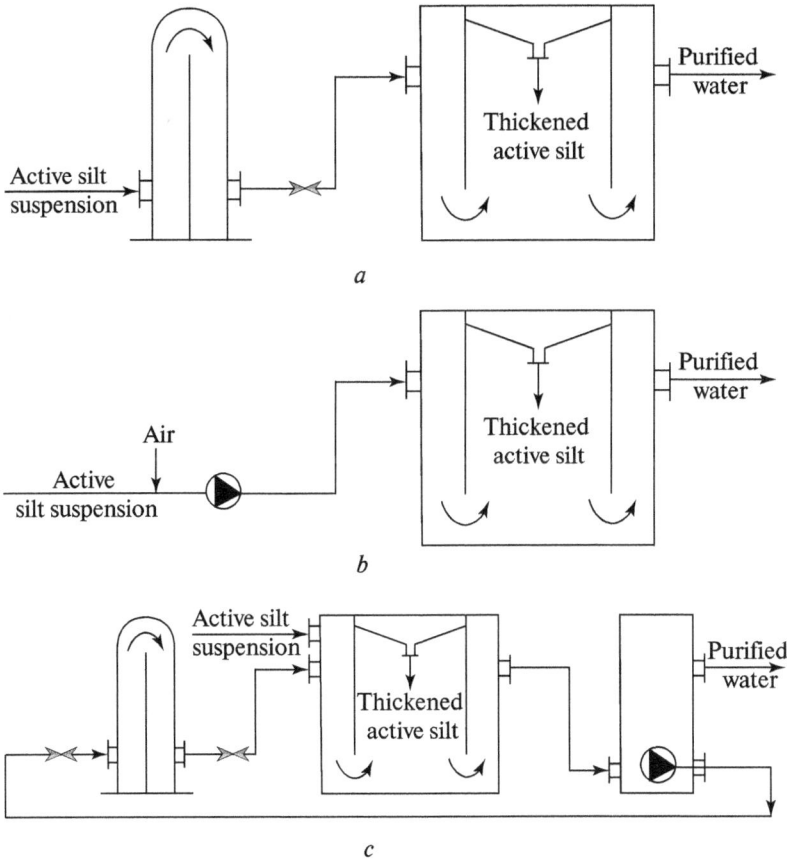

Fig. 8.1. Technological schemes of flotation thickening
of suspension of microorganisms:
a — pressure variant (with saturator) without working liquid;
b — without saturator with the use of pneumo-hydraulic system of
aeration; *c* — pressure-head variant (with saturator) with working
liquid

All variants of flotation thickening of active silt suspension, shown on Fig. 8.1, are embodied in practice, and choice of this or that approach depends on particular conditions and tasks being settled. In the case when after flotation thickening, active silt suspension is given on filter-extruder for dehydration till residual humidity of about 60–70%, variant *b*, Fig. 8.1 can be applied. At the supply of silt suspension on a press-filter, concentration of microorganisms in foam product in the limits of about 3–5% doesn't matter much, though expenditure of silt suspension, given on filter-press, will change. However, in the way of high rate of silt suspension, it can lead to additional costs and this, sure must be taken into account.

The impact of way of microorganism suspension saturation with air on its concentration in foam product after flotation is presented on Fig. 8.2. Presented averaged data testify some advantage of the way of active silt suspension saturation with saturator.

It should be noted that pressure drastic drop proceeds in the process of pressure-head flotation that can lead to the death of cells of microorganisms. To check this supposition, special research has been pursued.

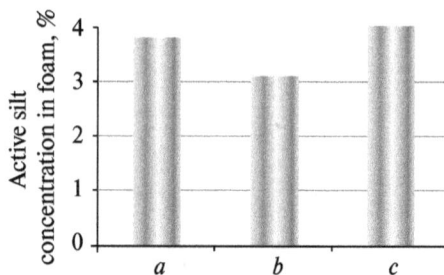

Fig. 8.2. Average values of active silt concentration in foam product after flotation thickening depending on the way of saturation with air of initial suspension of active silt (variants: *a*, *b*, *c* — look Fig. 8.1)

Experiment series have been made in the course of the research, which aim has been to define regimes whereon decompression impact can cause the largest inactivation effect on cells of yeast *Candida scottii*, and bacteria *Pseudomonas putida* which are often observed in the content of active silts of biotechnological productions, including hydrolysis plants.

In particular, there have been estimated the results of research on the different regimes of air saturation pressure and pressure drop speed. Sampling of 5 samples after 2, 5, 15, 20 and 30 minutes have been pursued and decompression treatment efficiency has been defined in the frames of each experiment.

Table 8.1. Regime and results of decompression treatment of yeast Candida scottii

Impact duration, min	15
Saturation pressure, Bar	14
Yeast suspension temperature, °C	20
Initial concentration of live microorganisms, (cells/ml)	10^8
Ending concentration of live microorganisms, (cells/ml)	10^3

Table 8.2. Regime and results of decompression treatment of yeast Candida putida

Impact duration, min	20
Saturation pressure, Bar	15
Bacteria suspension temperature, °C	20
Initial concentration of live microorganisms, (cells/ml)	10^7
Ending concentration of live microorganisms, (cells/ml)	10^2

Table 8.3. Impact of thermo-reagent processing of active silt suspension on concentration of polysaccharides in suspension liquid phase (active silt concentration in suspension is 0.5 %)

N	Processing parameters			Polysaccharide concentration, mg/l	Defecating efficiency (%)
	t, °C	pH	Delay, min		Microorganism concentration in liquid phase (cells/ml)
1 (control)	20	7.2 (init.)	—	11	$54.3/10^5$
2	40	7.2	5	35	$62.8/10^4$
3	80	7.2	5	52	$73.1/10^4$
4	85	7.2	5	54	$75.8/10^4$
5	90	7.2	5	53	$74.6/10^4$
6	20	8.0	5	9	$52.3/10^4$
7	20	8.5	5	9	$51.7/10^4$
8	85	8.5	5	8	$50.9/10^4$
9	85	8.5	10	35	$61.6/10^3$
10	85	8.5	30	31	$60.7/10^2$
11	85	8.5	45	31	$61.1/10^2$
12	85	8.5	60	9	$50.8/10^1$

It has been established the decrease in live microorganism concentration by several orders during decompression processing of suspension of yeasts *Candida scottii* for 5 to 15 minutes.

Table 8.2 shows the change in treatment efficiency of *Pseudomonas putida* bacteria for their inactivation with the use of preliminary optimal regime.

It has been established the decrease in live microorganism concentration by several orders during decompression processing of suspension of bacteria *Pseudomonas putida* for 5 to 15 minutes.

Together with inactivation of biomass of pointed microorganisms, the experiments on biomass inactivation and thickening with the use of thermo-reagent treatment have been held.

It's known that the intensification of the process of flotation thickening for active silt suspension depends on its preliminary treatment, for instance, with the use of chemical reagents, various physical impacts, of heating, in particular. During physical-chemical processing of active silt suspension, the processes of microorganism aggregation proceed, in particular, under the effect of polysaccharides distinguished in the process of such treatment. Table 8.3 shows values of polysaccharide concentration in active silt suspension, heated till 85 °C, being in liquid phase. It is observed whereat that polysaccharide concentration grows 3−5 times at pH raise till 8.5 and meanwhile, aggregation effect is observed visually.

The obtained data point on expediency of such physical-chemical treatment as one of the methods of the intensification of active silt flotation thickening and the inactivation of microorganisms in defecated liquid phase of the suspension.

Other way of inactivation and thickening of biomass by pressure-head flotation is the usage of ozone as a gas. Ozone flotation application in the pressure-head variant doesn't give advantages by thickening indicators in comparison with the air variant. Silt defecated water analysis shows that live microorganisms are factually utterly absent therein. However, costs for ozone-flotation of active silt suspension significantly exceed analogous indicators with the use of the pressure-head flotation air variant.

The presented examples point on the possibility to obtain thickened biomass of active silt or yeasts with simultaneous partial inactivation of microorganism cells as well as of defecated liquid phase with inactivation indicators, whereby in the series of ways <, the dropping of such waters into an open pond is permitted. It's worth to note that some from offered approaches, in particular, with the use of thermo-reagent processing, are expensive in the row of cases, for example, when there are no vapor surpluses on a plant. Yet, in the case of the usage of pressure-head flotation with the big value of pressure of air saturation in saturator, offered method can be rather competitive in comparison with the application of known technologies.

In individual cases, there can be also efficient the way, developed by the author, of active silt thickening with two working liquids saturated with the gas of different solubility in water, for example, with air and carbon [16, 25].

8.2 ELECTRO-FLOTATION AND NON-SEPARATIONAL METHODS OF MICROORGANISM BIOMASS EXTRACTION

The extraction of microorganism biomass by flotation methods, including electro-flotation ones, is simpler and cheaper than separational. We have pursued the research on the selection of flotation method for yeast biomass distinguishing, in particular, on electro-flotation extraction of yeast biomass.

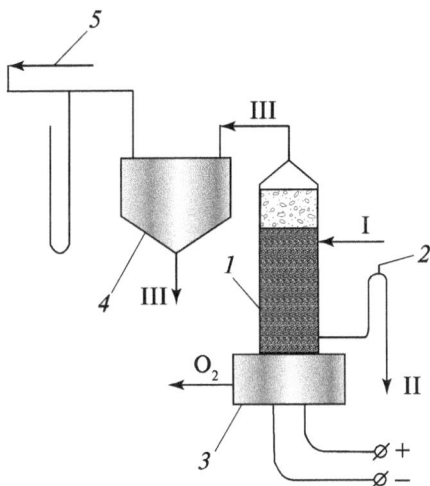

Fig. 8.3. Laboratory electro-flotation device scheme. Initial liquid is given into electro-flotation skimmer central part (*1*), bend — from lower part via hydro-seal (*2*). Gas bubbles are formed in electrolysis node (*3*). Electro-flotation skimmer upper part is filled with foam, which under vacuum impact (20–30 mm of mercury) (*5*) — vacuum-pump, is bent into foam collector (*4*). I — initial liquid; II — purified water; III — foam product

The scheme of laboratory installation for biomass electro-flotation extraction is shown on Fig. 8.3.

Two variants of electrolysis node are proposed (Fig. 8.4):

Cathode and anode space are divided by diaphragm (*7*, look Fig. 8.4) in the two cases. Anode space is filled with electrolyte (NaOH) solution for resistance decrease. Division between electrode spaces prevents the creation of explosive mixture of hydrogen and oxygen in electro-flotation skimmer and makes it possible to use gases separately. Experimental data have been obtained on above-described lab installation. Fig. 8.5 shows dependence of concentration of the liquid biomass, exiting flotation skimmer, from the liquid time-being in apparatus.

For to achieve liquid full cleaning from biomass in the installation given variant, the time of about 10–12 min is needed. Obtained data estimation has shown that electro-flotation speed exceeds regular flotation speed by about 2 times.

Fig. 8.6 demonstrates dependence of the biomass concentration in foam from the time.

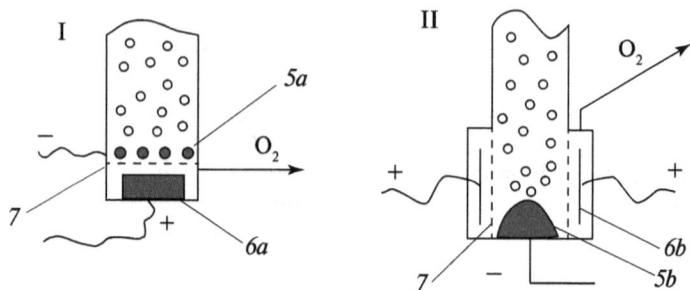

Fig. 8.4. Electrode node design:
I — with cathode horizontal positioning (*5a*),
accomplished in the form of net from steel 3 mm rods.
Anode (*6a*) — from graphite; II — vertical nickeliferous
anode (*6b*) and horizontal graphitic cathode (*5b*)

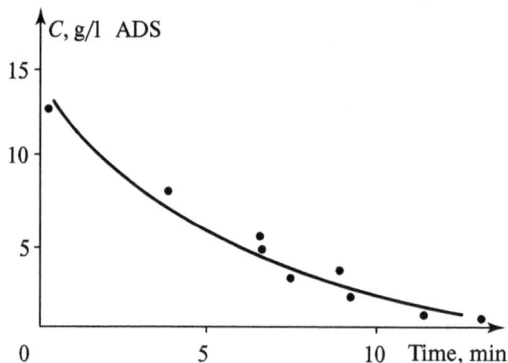

Fig. 8.5. Dependence of biomass concentration on exit from electro-flotation skimmer versus time

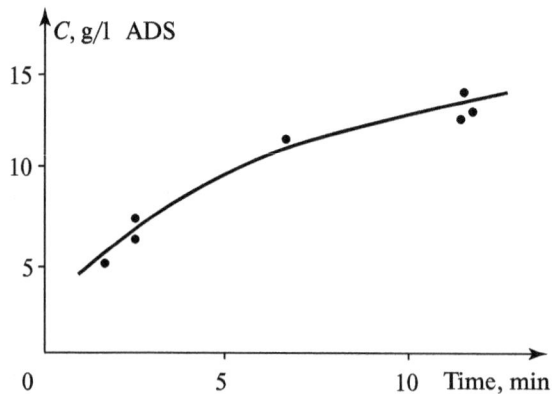

Fig. 8.6. Dependence of biomass concentration in foam versus electro-flotation time

Data, presented on Fig. 8.6, show that after more than 10 minutes, microorganism concentration in foam becomes practically constant and biomass extraction process can be ceased whereat.

To intensify electro-flotation process, an apparatus of original construction (Fig. 8.7) has been developed [31], it includes Shell *1*, which exterior side on, Sleeve *2* is fixed up for the supply of being defecated liquid part and reagent solution, Sleeve *3* for being defecated water major part supply, foam Gutter *4* with Sleeve *5* for foam product bend, Sleeve *8* for defecated liquid withdrawal, and half-supplying Walls *6* and *15* inside the shell, device with bringing horse (Pipeline *9*) for being defecated liquid level regulation, Filter *10* with grainy load, Electrode major pair *13* (anode) and *14* (cathode) and Electrode additional Pair *17*, plugged in continuous or alternate electric current source. For to hold Shell *1* in horizontal position, special Stays *16* are used.

Working principle of the offered electro-flotation apparatus is in the following.

Main part of initial wastewater is supplied via Sleeve *3*, and the smallest part of initial wastewater with reagent solution is supplied via Sleeve *2* and then, is processed in the zone of impact of continuous or alternate electric field in space between Electrodes *17*. Further, main and assisting flows mix in location zone of Electrodes *13* (anode) and *14* (cathode), plugged in continuous current source. In this zone, it proceeds the formation of flotocomplexes (pollution particles)-(bubbles of electrolysis gases (oxygen and hydrogen)) and their separation from being defecated liquid by the way of their surface emerging and bending via foam Gutter *4* and then via Sleeve *5*. Small flotocomplexes, that are late to emerge on surface, are being removed at being defecated liquid entrance into Filter *10*. Meanwhile, small flotocomplexes before entrance of water into filter coalesce, combining into large ones, which quickly surface-emerge upwards and separate out the being defecated liquid. Cleaned water along Pipeline *9* via water regulating Device *7* is withdrawn from the apparatus via Sleeve *8*.

Water cleaning efficiency in the proposed electro-flotation apparatus constitutes 93–99%, and 89–90% — in analog, with, correspondingly, spec. rates of electro-energy of 0.8–1.4 in the proposed — and of 2.1–4.9 in the known one.

Thus, the use of the proposed electro-flotation apparatus allows essentially to raise water purification efficiency at smaller spec. expenditures.

Promising way of wastewater purification is also the use of an ejection flotation. With the purpose to intensify the process and the achievement of more stable results, a chamber flotation device with ejection flotation skimmer has been developed, its scheme is shown on Fig. 8.8.

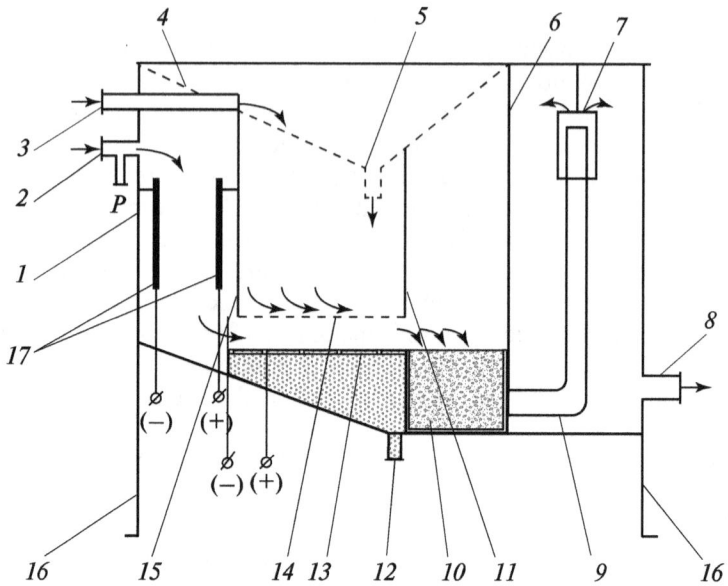

Fig. 8.7. Electro-flotation apparatus scheme

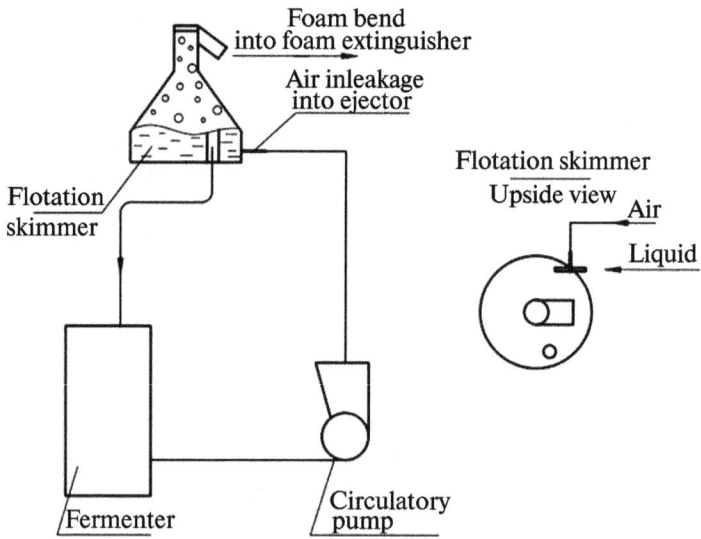

Foam bend into foam extinguisher

Air inleakage into ejector

Flotation skimmer

Flotation skimmer
Upside view

Air

Liquid

Fermenter

Circulatory pump

Fig. 8.8. Flotation device scheme

Liquid circulation is made via flotation skimmer lower part. Foam is withdrawn via decanted device in flotation skimmer upper part. Such flotation apparatus type advantage is the formation of very small air bubbles and low spec. costs.

However, it can be regarded to advantages of electro-flotation apparatus the possibility of additional disinfection of being cleaned liquids that is rather important at dropping of cleaned wastewaters into an open pond.

8.3 PROBLEMS OF EXTRACTION
OF METALS FROM TECHNOGENIC WATERS

To extract ions of non-ferrous metals from strongly dilute solutions, mainly the following physical-chemical processes are used: sorption, extraction, ionic exchange and others [63–71]. Let consider major virtues and drawbacks of pointed methods. Methods of sorption, ionic sorption and extraction are widely used in industry, as they have a series of widely-accepted advantages such as method embodiment simplicity, reagent availability, process regulation incomplexity.

Nevertheless, essential drawback, taking into account afore-pointed tendency to the increase of treated waters volume and decrease of distinguished component concentrations, is a rather low process kinetics. Indeed, for to realize ionic exchange, the time, needed for solution contact with ion exchanger, is about 15 minutes. During metal ion extraction by liquid ion exchangers, balance is being established only for 15–20 seconds, but 5–10 minutes are needed for lamination of organic and aqua phases. Besides, organic phase losses as a result of dissolution and entrainment are significant that in the way of essential volumes of poor solutes unpleasantly impacts process economy. Meanwhile, pollution of raffinates with organic substances much complicates ecological task settlement.

Tendency in hydro-metallurgic branch to the increase of processed solution volumes and the decrease of extracted component concentration demand the improvement of ionic exchange kinetic characteristics and the lowering of organic phase losses at liquid extraction. Modern paces of production demand maximal speed of extraction process, its largest efficiency, environmental friendliness, minimal losses and costs. From this point of view, method of ionic flotation represents scientific and practical interest, the method is based on specific properties that are characteristic for interface of phases. Surfactants, brought into solution, come into interaction with being extracted ions and molecules, adsorb on gas bubble surface and form a foam or scum on solution surface, enriched with being distinguished substance.

Ionic flotation method possesses the row of evident advantages against other extraction approaches: high productivity, efficiency at metal low concentration in a solution (from share of milligram to hundreds of milligrams in a liter) and, meanwhile, organic reagent losses do not exceed several milligrams in a liter at optimal chosen reagent regime. Method doesn't demand large material expenditures on realization, and its application is rather simple.

Thus, ionic flotation method is seemed to be efficient for purification and deep fine-purification of waste and recursive waters as well for precious metal extraction in industrial production and in hydro-metallurgy.

The ionic flotation can be applied as a major method for wastewater industrial cleaning as well as round with widely known methods. The ionic flotation application after existing on enterprises method of pollution removal provides for deep cleaning of industrial sinks and such that latter in the case series can be repeatedly used in production. Extraction of metals, lost together with industrial sinks, allows to improve mining raw use and to lower harmful emissions into natural environment.

Nevertheless, method hasn't won needed popularity what with the uncertainty of the methodology of flotoreagent-collector choice. For maximally efficient extraction of metal ions with the help of this method, it's necessary to choose surfactant correctly. At the moment, standard mechanism of flotoreagent-collector choice for each individual case and ionic flotation conditions haven't been approved. From time, choice methodology becomes more clear, but, nevertheless, it is not structured enough and too difficult to allow easy performing of surfactant choice and to widely ionic flotation process in industry.

Analyzing the matter of ionic flotation, one can emphasize 3 main periods. First experiments, performed at C19[th] end, were based on foaming and barbotage of individual bubbles for simple and comfortable formation and separation of liquid surface layers on the border with gas. The possibility of these methods usage for component separation wasn't perceived or it wasn't appreciated enough.

In 1903–1938, the second stage of the method development, connected with foaming application for component separation of real water solutions as well as of hydrosoles and micro-dispersions, began. In this period works, ion and molecule flotation were based on surficial activity of colligand itself.

Application of collectors for the extraction of surficial-active and weak surficial-active colligands is characteristic for the third stage. We can regard to beginning the year 1959 when Felix Sebba (South African Republic) published his basic work on ionic flotation [27] where there were presented the bases of

158

ionic flotation for the extraction of useful components from dilute solutions, including those for water purification. With the help of this approach some ions, molecules, fine-dispersed sediments and colloid particles interact with flotoreagents-collectors and are extracted by gas bubbles into foam or film on solution surface. The approach is promising for the conversion of industrial effluents, different technogenic waters.

The process of ionic flotation according to scheme is similar to classical flotation. Regular flotation principles can be spread on particle separation for those, which have ionic-molecular dispersity, by the way of the addition of surfactants which interact with extracted component ions and adsorb on gas bubble surface and dash out on solution surface.

It's worth to note in addition that ionic flotation differences and, for instance, mineral flotation ones are rather essential.

Mineral flotation is applied for the distinguishing of solid particles, suspended in water media, with the help of physical-chemical nature change of particle surface by surface-active reagent. The surface on definite sites becomes hydrophobic. As a result, air bubbles stick to particles and bring them on up surface. Being extracted substance during ionic flotation is in solution in the form of ions or colloid particles. During change of chemical conditions and applicable collector addition which should carry definite charge, solute becomes a product, having hydrophobic sites, with the help of which it sticks to bubbles and then lifts up to surface. Meanwhile, foam is being formed which represents concentrated form of initial substance. Hydrophobic product can in some cases give visible insoluble phase before it gathers in bubbles. But in some cases, insolubility is embodied on liquid-gas interface and becomes noticeable only at foam destruction and scum formation.

During mineral flotation, ending product consists of same solid particles as at the process beginning that are slightly changed only from surface. Ending product in ionic flotation has other chemical content in comparison with ions which were concentrated initially. It's worth to note that the flotation of minerals is usually held after ore crushing, and ionic flotation can be applied after leaching stage, though this leaching can be also a natural process, continuing for the very long time.

Necessary for particle separation differences at ionic flotation can be created artificially or increased with the help of special flotoreagents: collectors and regulators, acting with definite selectivity degree. Bubble stability, necessary for flotation process pursuing, can be supplied with reagents-(foam-formers), but collector itself possesses this property most often.

Collectors create or increase bubble capability to concentrate on bubble surface. In the cases majority, there are organic compounds of polar-apolar composition which are being fixed on particle surface and strengthen its hydrophobicity. Polar group of similar collectors interacts with bubble surface and causes reagent fixing, and apolar group is a water repellent and includes usually one or several carbohydrate radicals. Collectors, having enough long carbohydrate radical, usually possess high surface activity and foam-forming properties.

Regulators (diverse by substance content) promote (activators) or prevent (depressors) collector activity.

Foaming agents are surfactants which're adsorbed on bubble surface, decrease their size and uplift speed, improve foam stability.

Colligand — is a ion or molecule, being extracted.

Sublat — is a chemically individual substance, which content in, colligand concentrates on bubble surface.

Scum — extracted substance concentrate, formed on foam or solution surface.

Depending on process flow conditions, being extracted component kind and collector character, the ionic flotation process can be divided on row of processes which basis on, common principle of adsorption- and adhesion-bubble separation lays. Golman has offered to pursue a rational nomenclature of these processes, the nomenclature takes into account the following parameters:

1. Phase wherein floating particles accumulate:
   - Foam liquid phase
   - Scum
   - Organic liquid (flotoextraction)
   - Water phase (bubble fractioning or foamless flotation)
2. Character of floating particles, i.e. particles which content in, colligand concentrates on lifting bubble surface:
   - Ions and molecules
   - Colligand sediment particles
   - Carrier particles
3. Collector usage for to create and raise floating particle capability to gather on surface of bubbles:
   - Collector isn't used, process flows only thanks to surface activity of colligand itself
   - Collector is used

Usually, the following classification of processes is applied:

Flotation	Flotation			
	A — foam	B — scum	C — flotoextraction	D — foamless
1 — adsorption a — without collector b — with collector	A1a A1b	B1a B1b	C1a C1b	D1a D1b
2 — of sediments: a — hydrophobic b — hydrofibrated	A2a A2b	B2a B2b	C2a C2b	D2a D2b
3 — with carrier a — hydrophobic b — hydrofibrated	A3a A3b	B3a B3b	C3a C3b	D3a D3b

From the point of view of industrial usage for dissolved substances extraction, in particular, metal ions, the biggest interest at present is in sediment film flotation (B2) which is based on surfactants capability to make cations of the majority of non-ferrous and rare metals to fall out into sediment in the form of hardly soluble compounds, having high hydrophobicity (or additionally hydrofibrated).

Foam flotation (A) in comparison with scum one (B) demands more stable foam. As a result, in variant (A), surfactant concentration in waste solution is higher and foam product output is higher.

Because of organic phase emulsification, floto-extraction (C) can be embodied only at rather low intensity of aeration that, following kinetic considerations, is unacceptable for industrial usage but doesn't have essential significance, for instance, at concentrating of small amounts of substances in analytical chemistry.

For foamless flotation (D), relatively low efficiency of separation is usually characteristic. The variant of this process (Da) is promising, apparently, for fine-purification of solutions from surfactants.

According to kinetic possibilities, adsorption flotation (1) has advantages over sediment flotation (2) only at rather low initial concentrations of colligands of less than $10-15$ mole/l. As at such concentrations of surfactants, a foam, possessing necessary stability for scum flotation, can't be usually obtained, it's feasible to use sorption flotation in industry only in foamless variant (D1), first of all, for fine-purification.

That concerns flotation with carrier (3), in consequence of relatively low sorption capacity of chemical sediments, which at the moment are used qua carriers most often, this process is promising mainly for trace amounts of substances, including metals from technogenic water.

Major factors, influencing ionic flotation process, are the following: collector (surfactant) rational choice, optimal acidity pH of solution for process effective flow, bubble size and aeration intensity, expenditure, flotochamber characteristics, solution temperature, aging effect record, presence of neutral salts, solution concentration, foam stability and others.

Let consider basic methodologies on definition of process major parameters, worked out initially in the time of F. Sebba and continued further by other scientists, by A.M. Golman, in particular [27].

Ionic flotation efficiency depends mainly on correct choice of surfactant (detergent) and values of the process chemical parameters. Development of rational methodology of reagent properties evaluation is a vital task from the point of view of the process industrial realization.

F. Sebba was the first who had developed common requirements for the collector: a stability at micelle formation, a selectivity and so on.

Now then, one of the most important requirements for surfactants — is absence of micelle formation.

Let resume briefly some facts which contribute to or hinder micelle formation.

Neutral salts strive to lower the critical concentration of micelle formation that is preconditioned not by desalting action but by zeta-potential decrease. In relation to that ionic flotation can be often applied in solutions of rather high ionic force, this factor shouldn't be ignored.

Micelle formation critical concentration decreases at length increase of surfactant chain. Each two additional OH_2 groups lower the critical concentration of micelle formation on about one quarter. From the other side, carbohydrate chain octopus leads to micelle formation critical mass increase. Hence, it's willingly to use shorter and more ramified surfactant which satisfies process other requirments.

At temperature decay, tendency to micelle formation grows. However, little changes have no big effect in ionic flotation.

A selectivity is one of the poorly studied problems of ion and molecule flotation. One should distinguish in general case at collector usage the selectivity of interaction collector-colligand, adsorption selectivity on bubble surface and general selectivity of separation process.

Selectivity of interaction collector-colligand depends on equilibrium ratios and kinetics of corresponding reaction. Adsorption selectivity represents wider notion and includes also surface activity and adsorption kinetics. Separation process common selectivity depends not only on adsorption selectivity but also on an such factors as draining and foam reflux, organic phase nature, scum solubility and so on.

Then Sebba recommends some common rules as a manual for surfactant choice: collector should be charged oppositely regarding of ion being extract-

162

ed and satisfy ionic flotation major principle, consisting of the fact that ending sublat should be insoluble or poorly soluble. It puts some limitation on collector choice. Some sublats are so poorly soluble that any collector, containing more than 8 of carbonic atoms in carbohydrate radical, will be applicable. However, in some cases, solubility of sublat, which is a metallic soap, is rather high and at foam destruction, it can dissolve again. As a result, if scum still occurs it's low-stable. Nevertheless, with carbohydrate chain growth, stable insoluble scum forms.

Sebba has offered to pursue surfactant choice experimentally for each individual case, taking into account afore-described recommendations.

Thus, it's seen that mainly surfactant properties at ionic flotation are estimated experimentally. Such empirical approach drawbacks are process laboriousness and lasting. Besides, even if promising from the view of their extraction surfactants with definite polar groups are chosen for particular colligand, the question on optimal length for each group and structure of carbohydrate radical stays open.

Kuzkin S.F. and Golman S.F. more fully developed and scientifically substantiated choice of collector.

Following these authors, collector during ion and molecule flotation should possess:
- surface activity
- capability to form compound with colligand
- selectivity
- foaming properties

For ion and molecule flotation, collectors, having high surface activity, are desirable; significant differences in composition between solution volume and solute-gas surface only under this condition can be. It's known that surface activity grows at transition to higher homologs, however, critical concentration of micelle formation and water-solubility lower simultaneously. As a result, for instance, among the most spread ionogenic surfactants, containing one polar group and one linear carbohydrate radical of $C_nH_{2(n+1)}$ type, those substances are usually used for flotation that have carbon atom number from 8−10 to 16−18. The sage of the substances with ramified carbohydrate chain, with several apolar radicals or polar groups, different nature and mutual position of these groups essentially allow widening nomenclature of collectors applicable for flotation.

To choose the collector for the ionic flotation it's necessary to consider also that this process ending product (scum) must have low solubility and that sublat solubility falls with the increase of collector apolar radical length. Thus, at scum flotation of magnesium cation with laurate-anion, unstable scum of magnesium laurate appears which can dissolve again. However, if palmitate is used instead of laurate, the process proceeds normally.

If collector apolar group is responsible for surface activity presence, polar group must provide for formation of compound with colligand. Nature of link between collector and colligand in sublat might be various: electrostatic attraction of oppositely charged ions, coordinating link etc.

In the cases when flotation flows on ion-molecular level, two mechanisms of collector effect are possible fundamentally:

- compound collector-colligand is formed in solution volume, and then is adsorbed on bubble surface;
- collector is adsorbed first of all which then attracts colligand, and compound collector-colligand is formed on surface.

Probability of this or that mechanism depends on that how fully the interaction between collector and colligand goes on in solute volume. Thus, for example, in the case of Coulomb attraction of oppositely charged ions of collector and colligand, sublat, according to the theory of electrolyte solutes, can't exist in diluted water solution in the form of kinetic independent unit. If it's such, only second mechanism is possible — collector ions adsorb on bubble surface and create electric field which in, colligand ions concentrate by way of counterions. If the compound of collector with colligand represents rather stable formation in aqua solution, for instance, chelate and its dissociation in considered conditions can be ignored, the first mechanism has to be major.

Surfactant selectivity matter is settled by the authors rather approximately with the use of simplified approach which doesn't take into account adsorption kinetics, sizes, structure and hydration of ions as well as specific interaction between collector, as surfactant with ion, and colligand ions. Thus, to explain the selectivity, which is observed while extraction of ion, having the same charge, we've managed only approximately and on the basis of experimental data.

Collector at ionic flotation in comparison with mineral flotation fulfills foaming functions also. Foam character depends not only on collector nature and concentration but also on solute composition, gas rate and other factors. Besides, the requirements on foam properties are not similar for various processes. That is why the question on applicability of one or another collector is not settled experimentally in each individual case.

Golman A.M. in his works [27] made the attempt to systemize surfactant choice methodology, composing chemical model of ionic flotation process and describing its mathematically on the basis of methods, applied at the calculations of ionic equilibriums in electrolyte solutes.

He had analyzed the major reactions, going on during ionic flotation: interaction of ions (surfactant-$RX^{\pm l}$) with (colligand-$A^{\pm m}$) and competing ion $B^{\pm m}$ (l, m, n — ion charges), ignoring side reaction and processes (co-sedimentation and so on), wherein the pointed groups can participate, as well as intermediate compounds formation.

First important item is a definition of surfactant rate. Golman bases on that two reactions proceed in solute: $mRX^{\pm l} + lA^{\pm m} = (RX_mA_l)$solute which equilibrium is characterized by multiplication of activity and (RX_mA_l)solute = $RX_mA_l\downarrow$. Equilibrium of two reactions is characterized by the constant of sublat instability and its molecular solubility.

The author considered then a matter on electrolyte impact on ionic flotation process because in industrial solutes, significant quantities of elelctrolites present usually. Golman considered the case of the presence of some competing ion $B\pm n$ which can form soluble as well as insoluble compound with surfactant that, respectively, impacts collector (surfactant) rate.

Basing on notions developed by him on the possibility to describe quantitively ionic flotation chemical mechanism as well as on linear relation of solute such parameters as a product of solubility L, molecular solubility S, instability constants of compound K against number of carbon atoms N in carbohydrate radical of surfactant ($-\lg L \approx \alpha + \beta N$; $-\lg S_M \approx \alpha_1 + \beta N$; $-\lg K \approx \alpha_2$; α, α_1, α_2, β — permanent coefficients). Golman offered methodology of preliminary assessment of collecting properties of surfactant homological sequences and optimal length of carbohydrate radical.

Compounds with being extracted and competing ions are synthesized for two items of the chosen homological group and L, S, K values are defined. The obtained data are used for linearity coefficients α, α_1, α_2, β search. Thus, we have data for L, S, K calculation for various values of N_m, using equations of expressions for L and for K via equilibrium concentrations of ions which develop into expressions, connecting values of initial concentration of ions and degree of their extraction ε during flotation.

Analyzing offered methodology, one can make a conclusion that given method gives an opportunity to assess in the first approximation the results of ionic flotation. Similar calculations allow to define optimal value of N and give preliminary assessment of given homology group suitability for set task settlement. Nevertheless, at positive result, it is necessary to check properties of the chosen by flotation experiments surfactant.

Qua drawbacks of proposed methodology, the following should be underlined.

First of all, in proposed method, recommendations are not given on how to use this or that homological group for colligand extraction at collector choice. It's necessary to work out definite classes of collectors as well as to classify by particular chosen parameters the ionic flotation processes and to compare groups of collectors of definite homological groups with particular groups of ion flotation conditions.

Second, the author described mathematically only the situation of presence of one competing ion in solute, probably, because of the complexity of mathematical expressions and impossibility of their settlement on the moment of research with the help of computers. However, it should be taken in account that in practice, in working solute, the presence of several competing ions and other impurities is possible. These factors should be considered because they certainly impact on surfactant expenditure. We can make a conclusion that it's necessary to describe mathematically common mechanisms of the process of interaction between a collector and a colligand, accounting for a big number of factors impacting the process: effluent initial composition, metal ion initial concentration, their existing chemical form in wastewaters, collector availability, possibility of effluent effective fine-purification from collector residual quantity in the case of a necessity and instrumental design of the process. It's needed to try to classify all possible scenarios of the process development and create universal standard methodology of surfactant collector choice.

Collector should have surface activity, sufficient foam-forming properties, capability to form low-soluble compound with colligand, availability or synthesis simplicity, comparably low cost and capability for regeneration.

Collector choice for metal ion concentrating depends not only on collector's properties and characteristics but also on wastewater content exposed to purification, on chemical behavior and form of metal being in effluents, their initial concentration. Namely this is the reason why before wastewater supply for purification, their thorough analysis should be held. To make it possible to exploit widely ionic flotation process in industrial scale it's necessary to describe distinctly kinetics of the process of metal ion extraction from solute for the purpose to calculate and correct industrial facilities of continuous cycle. Presently, to describe flotation process, Ksenofontov B.S.'s multistage model [50] is used more and more often, the model's principles are necessary to account for at ionic flotation kinetics description. Theoretical basis for flotation process consideration as of multistage one consists of the following. On the first stage (state A), particle contacting with gas bubble proceeds. The process is reversable and, hence, particle sticking with bubble (state B) can happen or particle and bubble diverge after momentarily contact. Flotocomplex particle-bubble forms in the process of adhesion, the complex emerges at surface on the expense of extrusive (Archimedean) forces. Emerged on surface, into liquid upper part, flotocomplexes particle-bubble form foam layer (state C). Meanwhile, transitions not only from state A to state B and then to state C are possible but also reverse transitions, accordingly, from state C to state B and further to state A.

Thus, we can resume that a process for industrial exploitation of ionic flotation process should be modelled, considering modern multistage flotation theory, imposing on it of ionic flotation specificities, i.e. it's shown above that existing mathematical descriptions of ionic flotation process kinetics are not vital and must meet improvement with account of up-to-date notions.

For to use ionic flotation method in industry the availability of distinct methodology of flotation apparatus calculation is needed.

A.M. Golman, basing on his mathematical model of ionic flotation process kinetics, analyzed impact of major physical parameters of ionic flotation in their mutual interaction. Nevertheless, as it was considered above, the process mathematical model demands improvement and, hence, the methodology of calculation of flotation device physical parameters must be refined.

It's connected with that presently there is no calculation standard methodology of ionic flotation apparatuses and the question of sorbent (surfactant) choice is rather open and demands refinements.

It's known that for to pursue ionic flotation process, surface-active ions (collector ions) are inputted into solution, the ions' charge is opposite by sign to charges of being concentrated ions — colligand ions (from Latin *colligere* — collect). Then, gas 1 is supplied to solution from downwards via dispersant 2, gas 1 forms air bubbles 4 (Fig. 8.9). Meanwhile, drastic increase of interface of phases proceeds. Bubbles adsorb surfactant together with oppositely charged ions of being extracted component and then form foam 5 on solute surface. Bubbles are destructed, and as a result from, scum 9 is being formed (low-soluble, hydrophobic solid product, floating on liquid surface), containing the given ions in concentrated form — sublat (from Latin *sublatum* — lifted). Scum removal is done by any mechanical method.

Ionic flotation has been developed applicably to ion extraction with the help of surface-active ions of opposite charge that has determined the flotation name. Though analogous process is possible in another cases, for in-

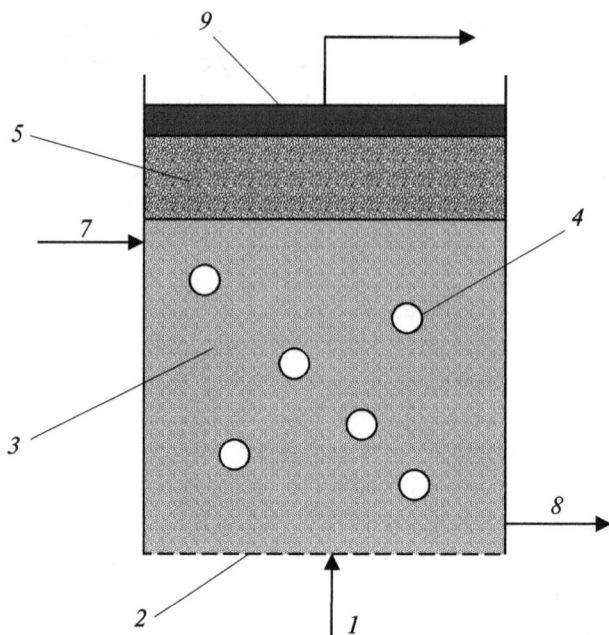

Fig. 8.9. Ionic flotation scheme:
1 — gas admission; *2* — gas dispersant; *3* — aqua solution;
4 — surface emerging bubbles; *5* — foam layer; *6* — extracted sublat
bend; *7* — initial solute; *8* — purified solute bend; *9* — scum

stance, at flotoextraction, foam separation or bubble fractioning, the ionic flotation is distinguished as separate kind of flotation division. Characteristic specificity of the given process is a scum forming that has served as basis for Kuzmin S.F. and Golman A.M. to call this process as a "scum" flotation.

Let consider in more details major processes, taking place at the ionic flotation of component from diluted solutes.

For to pursue ionic flotation, some substances are introduced preliminary into solute as a rule. There are collectors and regulators at its major.

Collectors create or increase the capability of particles to concentrate on bubbles' surface. In the majority of cases, it's a highly surface-active organic compound of polar-apolar (diphilic) composition (Fig. 8.10).

The collector with its polar group attracts electrostatically colligand to bubbles' surface or forms with it coordinating one or any other surface-active compound. Apolar group is a water repellent and includes usually one or several carbohydrate radicals.

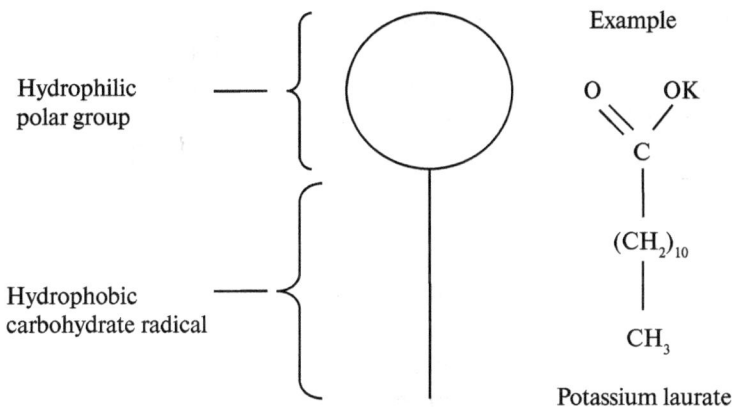

Fig. 8.10. Collector molecule diphilic composition

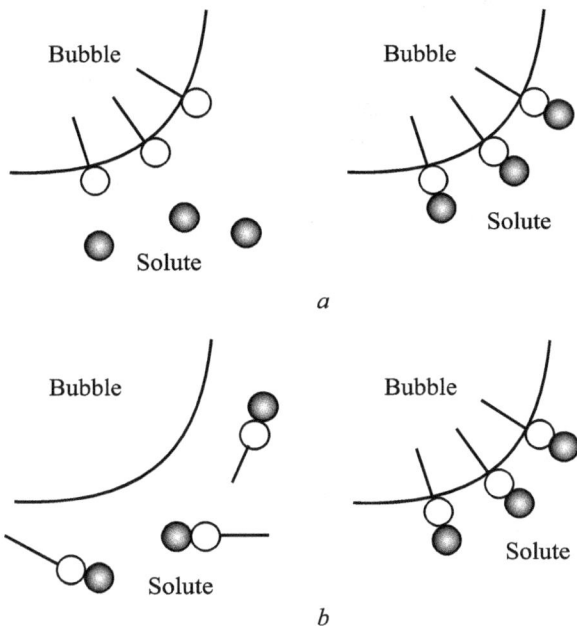

Fig. 8.11. Process of concentrating of ions from solute:
a — collector from solute is adsorbed on gas bubble surface and then attracts colligand ions; *b* — compound collector-colligand is formed in solute volume, whereat sublat, consisting of colligand and collector, is concentrated on bubble's surface

Two mechanisms of collector activity are possible:

- compound collector-colligand is formed in solute volume and then is adsorbed on bubbles' surface (Fig. 8.11, *a*);
- first of all, collector is adsorbed which then attracts colligand, and compound collector-colligand is formed on surface (Fig. 8.11, *b*).

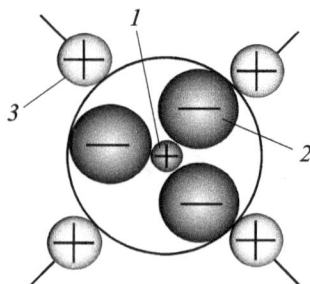

Fig. 8.12. Ions CO_3^{2-} are connecting link between ion UO^{2+}_2 and cations of tetrasubstituted alkylammonium collector: *1* — UO_2^{2+}; *2* — CO_3^{2-}; *3* — $R(R_2)N^+$

Probability of this or that mechanism depends on that how fully the interaction between collector and colligand proceeds in solute volume. Thus, for example, sublat, in the case of purely coulomb attraction of oppositely charged ions of colligand and collector, according to notions of electrolytes solute theory, can't exist in diluted (by colligand and sublat) aqua solution in the form of independent aggregate. If it's such, only second mechanism is possible — collector ions are adsorbed on bubbles' surface and create electric field where colligand ions are concentrated qua counterions. If compound of collector with colligand presents rather stable formation, chelate, for instance, and its dissociation in considered conditions can be neglected, the first mechanism should be the major one.

Regulators — are substances which provide for (activators) or hinder (depressors) collector's activity. Regulators are not always used, however, their usage in some cases appears necessary.

Let consider an example of double regulators' action. Uranyl cation UO^{2+}_2, which, sure, is not floated by cationic collector of cetyltrimethylammonium bromide type $[C_{16}H_{33}(CH_3)_3N]^+Br^-$, is present in the solute. However, if to add Na_2CO_3 into solution, uranium will occur in the solute in the form of urantricarbonate anion $[UO_2(CO_3)_3]^{4-}$ and the latter can be extracted with the same collector in sublat content $[C_{16}H_{33}(CH_3)_3N]^+_4[UO_2(CO_3)_3]^{4-}$ (Fig. 8.12). Thus, cations CO_3^{2-} play a role of activators.

For to describe the phenomenon proceeding in the ionic flotation processes, Golman A.M. and Kuzkin S.F. have offered mathematical model [27]. It represents one of the first attempts to create a common model, which:

- considers from common grounds a ion and molecule flotation, a sediment flotation and a flotation with carrier;
- grasps processes of flotoextraction, scum flotation, foam separation and bubble fractioning;
- is fair not only for stationary but also for non-stationary regimes;
- accounts for adsorption (adhesion) kinetics and impact of bubble mutual movement and of solution (suspension) on the process of adsorption (adhesion).

The given model is built on the consideration of flotoextraction continuous processes, scum flotation and foam separation. It's suspected that task of each of this is the extraction of some chemical element, for instance, metal *Me*. Whereupon, the following assumptions are accepted:

1) solute in apparatus is mixed ideally;
2) volume rate of initial product q and solute volume V in apparatus are constant;
3) bubbles are formed immediately on dispersant;
4) gas volume rate is constant;
5) surface-emerging bubbles have spherical shape and do not coalesce;
6) all bubbles are accepted of same diameter d at each solute composition;
7) adsorption value is equal to zero at the moment of bubble detachment from dispersant;
8) time of surface-emerging τ and adsorption dependence from the time of surface-emerging are also same at each given solute composition for bubbles same by size.

Main differential equation of the model has the form (8.1):

$$\frac{(dC_t^l)}{dt} = 1/V(C_t^i q - C_t^l q - k_{st} k_\tau k_s k_P C_t^l Q), \qquad (8.1)$$

where C_t^i and C_t^l — *Me* quantities in volume unit of coming (initial) and exiting from apparatus (low) products in arbitrary time moment t, respectively;

k_{st} — coefficient, characterizing adsorption stationary state on surface-emerging bubble (depends on solute composition and bubble diameter);

k_τ — coefficient, characterizing adsorption kinetics on surface-emerging bubble;

k_s — spec. (versus gas volume unit) surface of bubbles, characteristic for solute at the moment t;

k_p — coefficient characterizing separation fullness of floated *Me* (adsorbed by bubbles and carried by them away to solution upper border).

Let consider some individual cases of equation (8.1):
1) Equation of periodical ($q = 0$) process:

$$\frac{(dC_t^l)}{dt} = -k_{st}k_\tau k_s k_P C_t^i Q / V. \tag{8.2}$$

2) Equation of stationary regime ($C_t^i = C^i = \text{const}$, $C_t^l = C^l = \text{const}$, $\dfrac{(dC_t^l)}{dt} = 0$) of continuous process:

$$\frac{C^l}{C^i} = \frac{1}{\left(1 + k_{st}k_\tau k_s k_P \dfrac{Q}{q}\right)}. \tag{8.3}$$

3) Case when multiplication product $k_{st}k_\tau k_s k_P = k$, doesn't depend on solute composition change proceeding in the process. As q, Q and V are constant values, so at $C_t^i = C^i = \text{const}$ and $k_{st}k_\tau k_s k_P = \text{const}$, equation (8.1) can be integrated.

Let denote

$$k_{st}k_\tau k_s k_P = k, \tag{8.4}$$

where k — dimensionless coefficient characterizing the process.

As a result of conversions for stationary portion process, we will obtain:

$$\frac{C_t^l}{C_0^l} = \exp\left(-\frac{t}{T_1}\right), \tag{8.5}$$

where C_0^l — Me quantity in the unit of volume, going out from apparatus at initial time moment $t_0 = 0$;

$$T = \frac{V}{kQ}.$$

As in the common case, coefficients k_{st}, k_τ, k_s, k_P are complex functions of solute content and of other parameters of being considered processes, the further development of model notions should be aimed towards pointed functions clarification. Two mutually connected possibilities exist here:
• theoretical compositions, based on physical-chemical analysis;
• statistical analysis of experimental data with the purpose to construct statistical models for each from considered coefficients.

The following factors influence ionic flotation efficiency:
1) collector choice;
2) pH;
3) bubble size;
4) barbotage speed;
5) flotation chamber characteristics;
6) temperature;
7) aging effects;
8) neutral salts;
9) solute concentration;
10) foam stability.

Collector choice is often one of the most serious problems of ionic flotation. It happens because of lack of evidences concerning extracted ion nature. However, some general rules exist which can direct us during collector choice.

First of all, substance for ionic flotation is needed, which in aqua solution, can form surface-active ions, hence, non-ionogenic substances of spirits, ester, ketone and phenol types fall off. Among other substances, those ones which have necessary properties for ionic flotation are demonstrated in Table 8.4 (R designates long-chain carbohydrate radical of C_nH_{2n+1} type where n — might be from 8–10 to 16–18 ("fatty" radical)).

Table 8.4. Major types of collectors

Anion collectors	
fatty carboxylate	$R - COO^-$
fatty sulfonate	$R - SO_3^-$
fatty sulfate	$R - OSO_3^-$
fatty benzol sulfonate	$R - C_6H_5 - SO_3^-$
fatty phosphate	$R - OPO_3^{2-}$
fatty α-sulpho-carboxylate	$R - CHSO_3^- - COO^-$
Cation collectors	
fatty primary amin	$R - NH_2$
fatty monosubstituted ammonium	$R - NH_3^+$
fatty disubstituted[1] ammonium	$R_1R_2 - NH_2^+$
fatty trisubstituted[1] ammonium	$R_1R_2R_3 - NH^+$
fatty tetrasubstituted[1] ammonium	$R_1R_2R_3R_4 - N^+$
fatty pyridinium	$R - C_5H_5N^+$

[1]In these cases, only one of the radicals should be fatty one.

The use of compounds with ramified carbohydrate chain, with several apolar radicals and polar groups, various nature and mutual disposition of these groups allow to widen much nomenclature of the collectors, applicable for flotation.

Meanwhile, collector should be charged oppositely towards being extracted ions. As the majority of metals, usually being in drains, represents cations, the application of such anion collector as laurate at pH = 8 should lead to purification from the majority of them. However, the most effective and economizing collector can be found for particular case only by experiment results. For example, for gold containing solution with gold cation content 2.2 mg/l at pH = 10, anionic surfactant — is a trialkylbenzylammonium chloride taken with large excess. In addition, 98% of gold from initial solute are extracted. From the other side, many among precious elements are in drains in anionic form, for example, complex silver in photographic industry and cuprum from galvanic baths. They can be concentrated with the use of cation collectors of fatty ammonium salt type.

Moreover, sometimes, non-charged molecules can be used. Thus, fatty amins fit for extraction of anions only from acidic solutes when they form fatty ammonium ions. And non-charged amin molecules, for example, can be exploited for the extraction of those cations which are capable to create coordinating bond with free pair of nitrogen electrons. Such ones are cuprum and nickel ions, in particular.

Sometimes, difficulties at collector choice are connected with colligand charge uncertainty. At anion high concentration, cation can form anionic complexes with many of them. That is why, uranium might be extracted in the form of anionic uranylsulphate complex on anionite or floated with the use of cation collector. Besides, it's been observed that freshly prepared solutes behave themselves in other way than old ones. It can be connected with the complex formation duration.

Ionic flotation major principle is in that ending sublat must be non-soluble or low-soluble in water. It imposes some limitations on collector choice, in particular, it's known that with apolar radical length increase of a collector, sublat solubility falls down. For instance, one of the cheapest and available surfactants — dodecylbenzosulfonate — not always can be applied because some of its soaps with heavy metals are too highly soluble. Thus, at the ionic flotation of magnesium cation with laurate-anion, unstable scum of magnesium laurate, which can dissolve again, occurs.

Potential collector important characteristic is a micelle formation critical concentration (MCC) value. Micelle represents a sphere, containing surfactant diphilic molecules which hydrophobic parts are aimed inside and form sphere's body, and hydrophilic groups are oriented outside and are placed in the contact with water (Fig. 8.13).

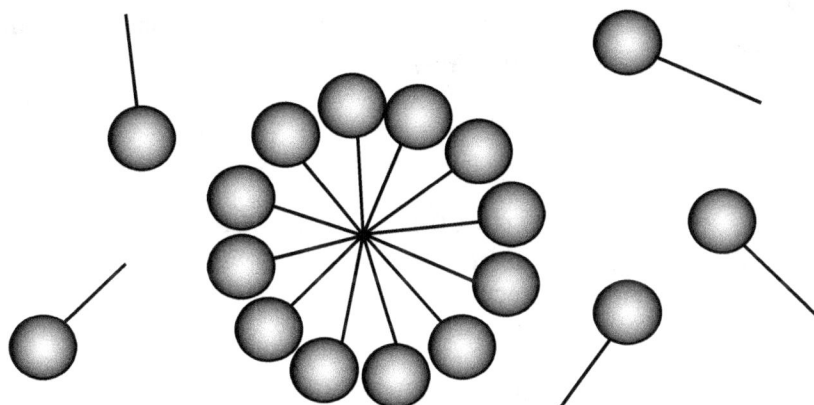

Fig. 8.13. Formation of spherical micelle of a surfactant

The matter is that the presence of micelles in ionic flotation is harmful as the presence of single surfactant ions is necessary when ion double nature can be exploited, i.e. the presence of charge on one terminal and of hydrophobic group — on another one. Thus, it's desirable to choose substances with maximal MCC to exclude micelle formation. MCC values are demonstrated below for some surfactants, in moles in Table 8.5.

Table 8.5. Critical concentration of micelle formation of some surfactants.

Name	Formula	MCC, mole/l
potassium laurate	$C_{11}H_{23}COOK$	0.02
potassium myristate	$C_{13}H_{27}COOK$	0.006
hydrochloric acid laurylamin	$C_{12}H_{25}NH_3Cl$	0.03
natrium laurylsulfate	$C_{12}H_{25}OSO_3Na$	0.006
natrium cetylsulfate	$C_{16}H_{33}OSO_3Na$	0.0004
natrium oleate	$C_8H_{17}CH = CH(CH_2)_7 COONa$	0.001

One should note that at carbohydrate chain elongation, MCC lessens. It's known that increase of carbohydrate chain octopus leads to MCC raise. Hence, it's willingly to use in the ionic flotation the shortest and the most ramified surfactant which will meet other requirements of the ionic flotation process. Those collectors seem rather interesting which have several polar groups capable to interact with colligand. Dodecylbenzyldiethylentriamintetraacetic acid (DBDTTA) can serve as an example:

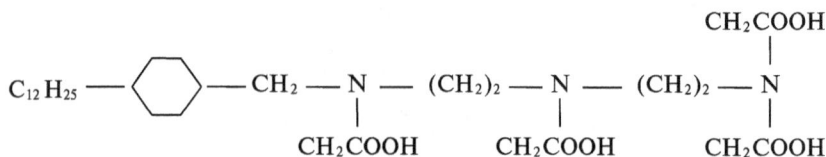

$$C_{12}H_{25} - \langle \ \rangle - CH_2 - N - (CH_2)_2 - N - (CH_2)_2 - N \begin{matrix} CH_2COOH \\ | \\ | \\ CH_2COOH \end{matrix}$$

with CH_2COOH groups on the N atoms: CH_2COOH, CH_2COOH, CH_2COOH.

This compound is a surface-active analog of well-known complexon — ethylendiamintetraacetic acid (EDTA). DBDTTA as well as EDTA form stable complexes with metal cations and is an effective collector for them.

One of the most important indicators — pH — can influence ionic flotation process by the way of change of sublat solubility, being extracted ion charge, colligand/collector ratio, foam stability and so on.

Not many anionic collectors exist that can be used at low pH. It's considered that the only applicable collector for acidic environment are fatty sulfates and fatty α-sulfoacids. Together, as was mentioned, fatty amin's application is possible in some special cases.

One can experience sometimes the substances having as cation as anionic active centers. In this case, depending on pH, they can be extracted both by cation or anionic collectors. Cuprum hydroxide, for instance, relates to substances that can be floated by cation as well as by anionic collector.

Let consider trivalent metal Me^{3+} which exists in acidic environment in the form of $Me(OH_2)_6^{3+}$ cation. At pH raise, so at concentration raise of OH^- ions, some protons of hydrating water molecules which're acidic by character, are attracted to OH^- ions and form water. Residual hydroxide interacts with hydroxyl of near metal ion, releasing water and forming oxygen bridge:

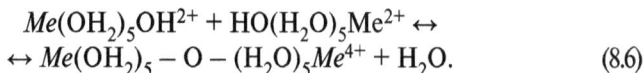

$$Me(OH_2)_5OH^{2+} + HO(H_2O)_5Me^{2+} \leftrightarrow$$
$$\leftrightarrow Me(OH_2)_5 - O - (H_2O)_5Me^{4+} + H_2O. \tag{8.6}$$

Oxygen bridges formation continues till the point when product, because of cross-compounds becomes insoluble and doesn't fall out into sediment in the form of hydroxide. Thus, this process is analogous to polymerization formally.

Ionic flotation practice has shown that number of collector ions necessary for given ion extraction is defined by its charge. For example, to extract ferrocyanide ion, having four negative charges, four single-charged ions of cation collector are needed. In the afore-considered example (8.6), initial hydrated ion has charge 3^+. It means that for its extraction, three single-charged anions of surface-active substance are needed. Hydrolysis first product with oxygen bridges has common charge 4^+ but ion contains two atoms of metal Me. Hence, only collector two anions instead of three are needed per one metal atom. Thus, stochiometric ratio improves. Meanwhile, with polymerization raise, this ratio becomes more favorable, and at the beginning of hydroxide sediment falling, needed relative quantity of collector decreases till the level, observed in the processes of metal flotation. So, for example, it has been shown that at pH = 6, for full extraction of cuprum 1 mole, only 0.01 mole of anion collector has been necessary. As at this pH value the sedimentation is noticeable before the flotation, term "sediment flotation" has been offered for such cases.

Thus, pH exists which flotation efficiency of given colligand is maximal at, and it can be used for selective flotation.

It's necessary to strive to bubbles' smallest sizes according to being developed by us flotation theory as a multistage process. In addition, it can be shown that required gas volume is proportional to bubble radius (Table 8.6).

Table 8.6. Nitrogen volume, necessary for 1 mM secondary natrium alkyl sulfates, depending on dispersant holes size

Capillary radius, mm	0.45	0.36	0.25
Nitrogen volume, l	375	315	191

However, bubbles reflection from surface should be excluded owing to insufficiency of their energy for penetration through surface. Given effect danger is in that emerging bubble can jostle away some sublat particles back into solution.

As at barbotage big increase the extraction falls down, it's necessary to maintain optimal speed, which at, bubbles calmly pass the surface. In addition, sublat reverse transition into solute volume is minimal.

A barbotage speed is needed to be regulated such that it would exceed a foam destruction speed only a little. Then foam layer of necessary thickness might be obtained. As a collector is extracted gradually, foam becomes more stable, so barbotage speed should increase some.

Chambers' various constructions are needed for various scales. First of all, if gas is a being introduced from outwards, a device (for instance, aerator in the form of pored wall) for bubble creation is necessary (Fig. 8.14).

It's desirable to input collector in small amounts, not touching foam layer, that is why it's comfortable to use capillary input hole situated some higher of pored disk. Also, input hole can be installed with the purpose to add initial solute or solution, regulating pH.

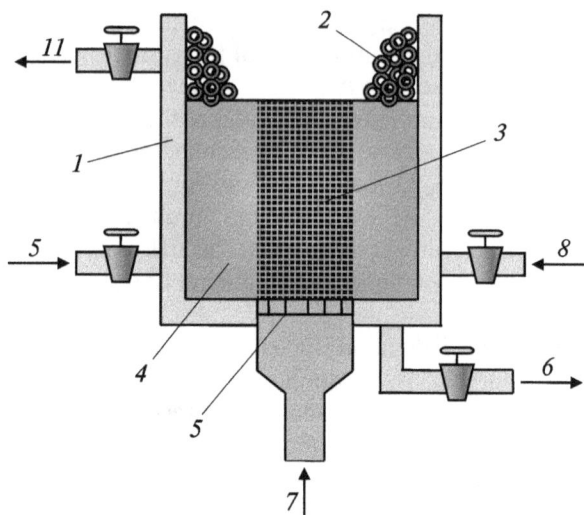

Fig. 8.14. Scheme of ionic flotation chamber:
1 — vessel with water-shirt; *2* — sublat; *3* — gas bubbles;
4 — solute; *5* — pored wall; *6* — bend of used solution via drain
faucet; *7* — compressed gas input; *8* — initial solute input;
9 — collector input; *10* — caught sublat bend

Chamber upper part should be much wider than filter diameter. Together, bubbles lift up almost vertically, not bumping walls and, that is even more important, are destructed in chamber center. And what about the foam, it's extruded into calm zone near chamber walls. As a result, it becomes less stable and sublat is gathered by chamber perimeter. The danger of sublat transition increases also. Fig. 8.14 shows chamber of such type.

At solute big quantities, a rectangular chamber is the most comfortable. It can be provided with devices for foam dragging through edge. Row of regulating walls will allow to pursue selective flotation. One can create continuous flowing system on that basis.

Temperature impact on ionic flotation is connected with effect upon foam character. It's expressed mainly when the temperature rises above sublat melting point. The latter is determined by the nature of being extracted ion and collector. Some sublats exist that melt below water boiling point. Meanwhile, liquid sublat drains to Plateau triangles instead of foam forming. In this case, it must be pumped out.

However, sometimes, melted sublat effect completely hinders foam formation. One should avoid similar situation. If high temperature at industrial process is inevitable it's necessary to change collector to another one which has higher melting temperature.

From the other side, micelle-formation tendency increases at temperature decreases. However, small change doesn't have big significance in the ionic flotation. Because as a result of long storage of a solute, a complex- and micelle-forming are possible, it's desirable to use freshly prepared solutes of a collector and exclude colligand aging.

Neutral salts impact on the process by one of the two directions. Firstly, they lower critical concentration of micelle-formation that's why their presence can cause forming of micelles.

Second effect is connected with influence on extracted ion which as a result, a complex-formation is possible. By this reason, complex-forming ions, for instance, chlorides and sulfates, should be by possibility excluded or collector of another sign should be taken. For example, cuprum's extracted usually in the form of cation with the help of anionic collector. However, if a little of chloride, for instance, natrium chloride, is present, it's much worth to extract cuprum as an anion, using cation collector.

Ionic flotation is effective in very highly diluted solutes. It's easy to extract ions at concentration of several milligrams per 1 l. The following characteristic concentrations for ionic flotation are usually said: $10^{-4}-10^{-3}$ M (not more than 10^{-2} M). Concentration upper limit is confined by complex-formation possibility. If flotation doesn't go on normally, solute can be impacted by a dilution.

Foam nature depends on pores size, barbotage speed, colligand and collector nature as well as on solute pH. Stable foams are desirable not only because they are difficult to be destructed but also because they stand larger quantities of initial solute that decreases efficiency in the case of colligands separation. If foam is stable, it can be destructed by hot air flow or by pulverization of ester sprays or of vapors. Yet, foam part is so stable that it's better to change flotation conditions such that less stable foam would be obtained. It's often enough to exploit big chambers or wider pores in dispersant for it.

To assess potential industrial collectors it's needed to formulate demands to them. The following requirements to collectors might be added to considered ones above:
- selectivity to target metal;
- process short durability (5–10 min);
- collector regeneration possibility;
- chemical stability;
- minimal entrainment with chamber solution (in the frames of maximum permissible concentrations — MPC);

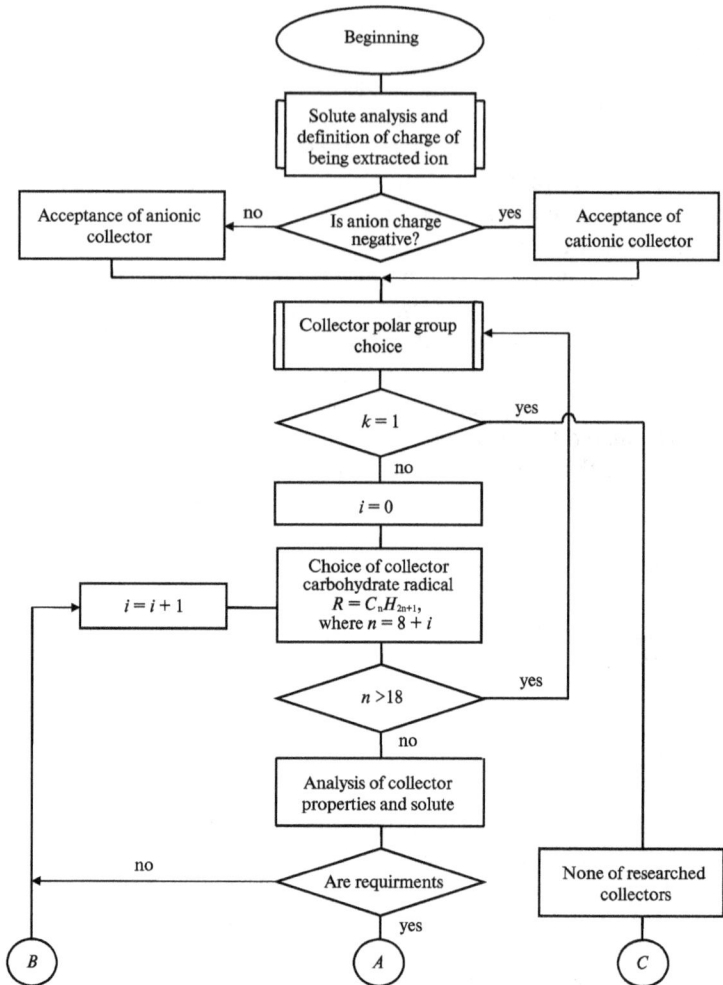

Fig. 8.15. Block-scheme of collector selection (beginning)

- low toxicity;
- flash high temperature;
- minimal rate versus extracted metal unit;
- low sensitivity to salt background.

As a result of the analysis on recommendations for collector choice [31] as well as being ruled by thoughts expounded above, it seems useful to formalize the given process. Let make algorithm for pursing of events that are necessary for collector correct choice (look Fig. 8.15–8.16).

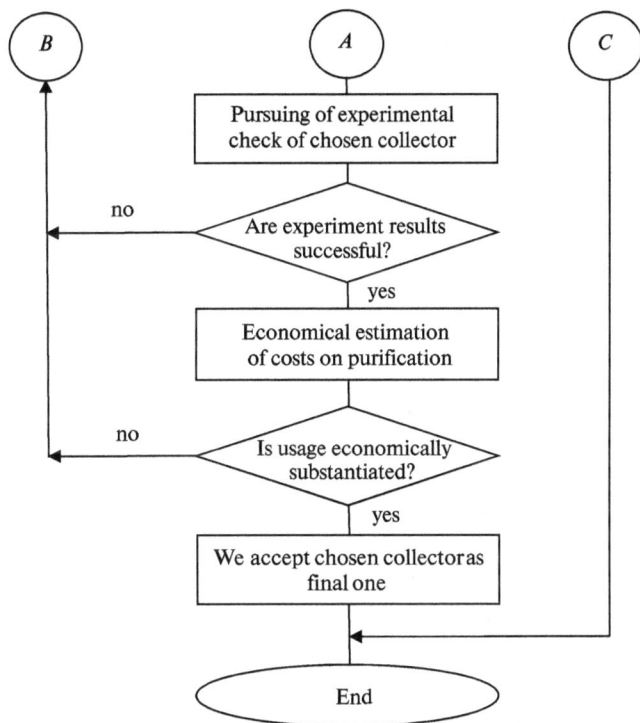

Fig. 8.16. Block-scheme of collector selection (ending)

It's needed to take into account that collector choice adds in also the selection of optimal conditions of the process flow as well as adds in some parameters of being cleaned solute if there is a necessity of their regulation.

The analysis of content and properties of original solute allows to define being extracted ion charge. Major stages of the analysis are presented in the form of procedure on Fig. 8.17.

Collector type is selected by this procedure. Meanwhile, it's worth to note that the scheme doesn't account for particular case of the use of non-charged collector, that was aforementioned about, as it's possible only for small number of ions at conditions special choice.

Collector selection is split into two parts: choice of polar group and carbohydrate radical length. Polar group is clarified in the separate procedure (look Fig 8.16). Radical various lengths for each chosen group in cycle are checked with radical consequent elongation because, as it's been already noted, the shortest radicals are preferable.

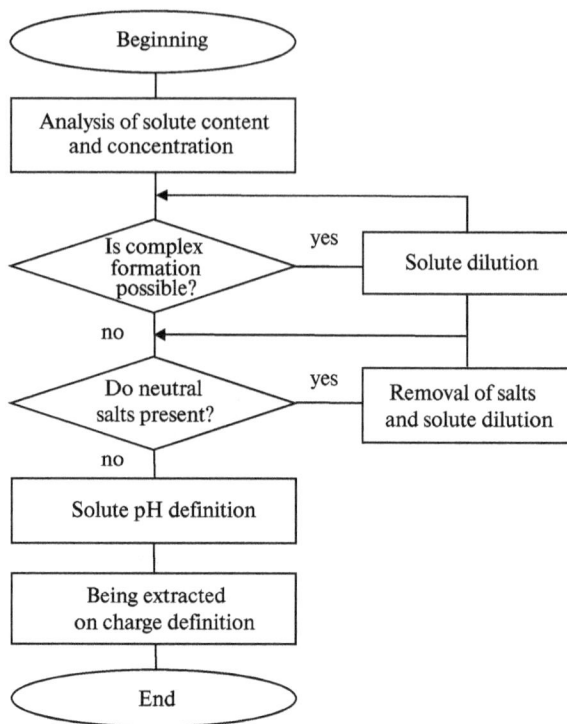

Fig. 8.17. Procedure of solute content analysis
and being extracted ion charge definition

Preliminary chosen collector analysis adds in check of accordance of its properties to the ionic flotation process demands (Fig. 8.15). Solute parameters are checked with collector critical parameters such as melting temperature of a collector and micelle-formation critical concentrations (MCC). In addition, as the temperature can cause influence on MCC, first of all, temperature is regulated and then, concentration is. Together, all of it is not a subject for check if it's been clarified immediately that a terminal sublat is well-soluble in water. In this case, it's left to try to change solute pH for solvability complication or to go to next trial ($i = i + 1$).

It's worth to note that the correction of solute parameters is far from being always possible and possible only in a narrow range. Industrial process conditions don't allow to influence strongly the parameters (for instance, temperature) sometimes, but still sometimes, the demands can be contradictory and one and the same parameter causes multiple activity. In particular, pH change doesn't influence not only sublat solubility but the charge sign of being extracted ion. Together, pH values aren't taken into account at collector's polar group choice.

Efficiency analysis of preliminary selected collector can be held also with the use of equations developed by Golman A.M. and Kackovsky I.A. Let's consider these relations.

Equation of interaction of collector $A^{\pm m}$ with colligand $MeX^{\pm l}$, deduced by Golman A.M., has the following form:

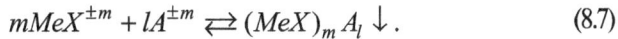

$$mMeX^{\pm m} + lA^{\pm m} \rightleftarrows (MeX)_m A_l \downarrow . \tag{8.7}$$

Expression has been deduced for (8.7) that looks like this in simplified way:

$$(EM)_{(R/E)} = (m/l)^m (R - E)^m (1 - E)^l C_A^{(m+l)}, \tag{8.8}$$

where EM — sublat solubility embodiment $(MeX)_m A_l$;
C_A — initial concentration of collector ion $A^{\pm m}$, mole/l;
E — degree of colligand extraction into sediment;
R — collector rate in shares from stoichiometry, whereupon

$$R = \frac{lC_{MeX}}{mC_A}, \tag{8.9}$$

where C_{MeX} and C_A — ions $MeX^{\pm l}$ and $A^{\pm m}$ initial concentrations.

Expression (8.8) allows to calculate $(EM)_{R/E}$, i.e. EM value that is necessary for achievement of extraction degree E upon given collector rate R.

Kackovsky I.A. and some other authors have shown that EM of sublats, formed by being extracted ion in homologous group of ionogenic collectors, can be calculated according to the equation:

$$\lg EM = a + bn , \tag{8.10}$$

where n — number of carbon atoms on surfactant radical;
a and b — constants for given surfactant group.

The use of equations (8.8) and (8.10) allows to prognose properties and necessary collectors rate and to lower essentially the number of flotation experiments at choice and testing of reagents.

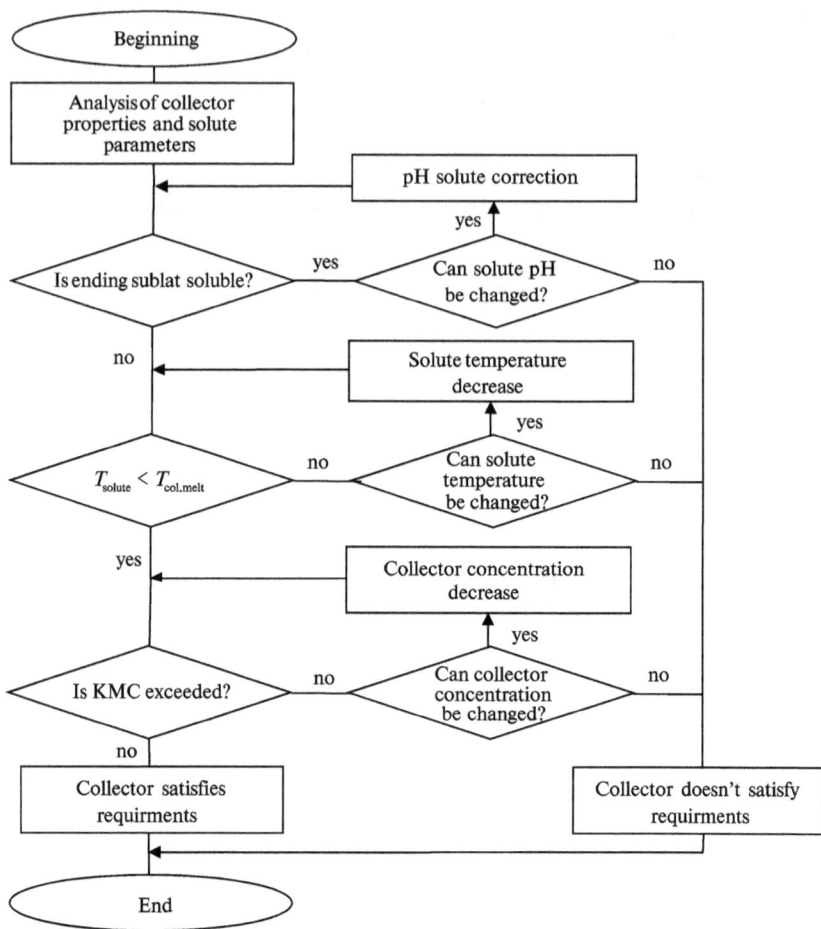

Fig. 8.18. Procedure of analysis of collector properties
and solute parameters

Returning to the consideration of algorithm of collector choice, we want to denote that the scheme also accounts for variant whereat none of investigated substances fits for given case qua a collector. The matter is that possible collector variants aren't the only from Table 8.4 list. It's the set of more or less researched compounds suitable for ionic flotation. In this regard, it's possible that substance with suitable properties for given concrete case will be clarified still or has more complex composition that the algorithm provides for.

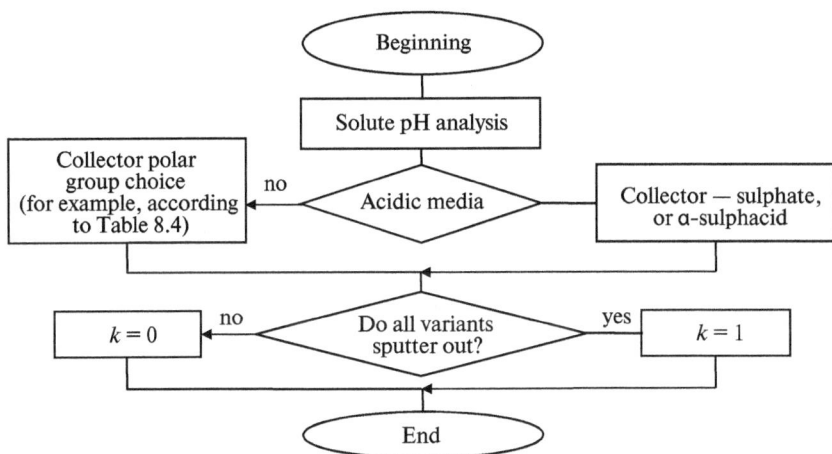

Fig. 8.19. Procedure of collector polar group choice

Part and parcel of the scheme of collector choice is an experimental check. It can be proved or a disproved by experimental way the possibility of applicability of selected collector in given conditions.

Collector estimation is held on final stage from economic point of view. Meanwhile, it's necessary to take into account collector required quantity versus set volume of being purified water, its price as well as price of extracted precious components if such ones exist. From the other side, the possibility to apply ionic flotation waste for marketable product manufacture exists. In this case, to estimate economical expedience of chosen collector application, one should consider additional costs on manufacture of marketable product and its market-value.

In the condition when all afore-expounded requirements to potential collector are fulfilled, one can regard this potential collector to final one.

Thus, collector choice algorithm is offered for ionic flotation pursuing. It includes some simplifications same as any scheme. The matter is that mutual connection of various properties of substances and process parameters is rather complex and isn't studied to the end that is why we haven't managed to automize the process completely. Besides, big variety of existing and again appearing compounds do not allow to account for all possible substances and their properties in the scheme.

From the other side, the algorithm doesn't give vivid notion about collector selection method in a whole, summarizing existing theoretical and experimental data. Besides, the proposed scheme might be used as a manual for compound properties study which applicability for ionic flotation isn't known.

185

Let's note that scheme revision and correction with account of new data, dealing with being considered sphere, is further possible.

Let show the example of collector choice for collective flotation of cuprum and iron from defecated solute obtained as a result of vitriolic leaching of Arizonan red copper.

Initial solute has pH = 2.6 and contains 0.5 g/l of cuprum and 0.08 g/l of iron. As it was aforesaid, acidic media is unwanted for ionic flotation because, in particular, it narrows collector possibility range. In this regard, pH is raised till 7.5. As in these conditions, being extracted components are in solute in the form of cations, an anionic collector is in demand. We have preliminary chosen natrium soaps of fatty acids qua a collector. As a result of consequent check of radicals of various lengths, we accept natrium laurate as satisfying all the requirements. It's been determined according to experiments results that mixture of several natrium fatty acids of coconut oil is the most effective. It is collector NeO-Fat 265.

Continuous flotation of iron and cuprum hydroxides was held in specially constructed pneumatic four-chamber flotomachine with 2 l capacity. At collector rate of order of 20 mg/l and the time-being in flotomachine of about 4 min, 97–99% iron and cuprum extraction was obtained.

According to data of literature sources [64–66. 68–70], let make a table, containing collector examples applicable for extraction of various ions with heavy metals.

Ethyl and hexyl xanthogenates — flotoreagents with common formula:

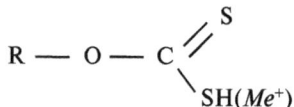

$$R - O - C \overset{\displaystyle /\!/ \, S}{\underset{\displaystyle \backslash \, SH(Me^+)}{}}$$

where R = C_2H_5, C_6H_{13}.

Thus, assortment of reagents for ionic flotation of heavy metals isn't very rich. Only alkyl carboxy reagents, naphthenic acids and, less, alkyl sulphates and alkyl sulfonates have technological application.

It's known that chelate-forming reagents are the most promising: they are the most selective, provide for solute deeper cleaning in one step than others. Besides, at correct dosing, they don't require waste polishing. Nevertheless, these collectors are more expensive, their application is feasible at collector regeneration or at waste polishing from heavy metals after reagent method usage.

The reserve for "new" collectors is, first of all, known classes of industrial extractants for non-ferrous and rare metals (phosphorus containing, β-diketones and others).

Table 8.7. Some collectors for heavy metal extraction

Ions	Collector
Be^{2+}	potassium stearate
Zn^{2+}	saponified fat, CA, NA, DAH, HCA, polyamins, chloride lauryl ammounium, ethyl and hexyl xanthogenates, EMKO*
Cr^{3+}	dodecylsulfate + dodecanate, natrium oleate, DAH, HCA, benzyl amin, EMKO
$Cr_2O_7^{2-}$	bromide cetyl ethyldimethyl ammonium
Fe^{2+}	KK
Fe^{3+}	saponified fat, NA, DAH, HCA, EMKO, NeO-Fat 265
Cd^{2+}	CA, NA, 2-pyridyl 1,2,3-triazine
Co^{2+}	CA, NA, dodecyl sulphate, DAH, HCA, ethyl and hexyl xanthogenates, benzylamin, saponified fat
Ni^{2+}	natrium and potassium, saponified fat, CA, HK, waxes, dodecyl sulfate, ethyl and hexyl xanthogenates, DAH, HCA, polyamins, fatty amins, EMKO
Pb^{2+}	2-pyridyl 1,2,3-triazine
Cu^{2+}	potassium and natrium laurate, dodecyl sulfate, natrium oleate, saponified fat, CA, NA, waxes, DAH, HCA, polyamins, fatty amines, ethyl and hexyl xanthogenates, 2-aminoethanol, NeO-Fat 265
Silver anionic complex	salts of fatty ammonium
Au^{3+}	trialkyl benzyl ammonium chloride
UO_2^{2+}	tetrasubstituted alkyl ammonium + natrium carbonate
Radioactive	alkyl benzol sulfonate, alkyl sulfate, fatty α-sulfa carbon acids, amino polycarboxy acids

*EMKO — abbreviation for special reagent.

Denotations used in Table 8.7:

saponified fat — alkyl carboxylate with radical $R = C_{14} - C_{16}$;

CA — vat acids, i.e. ramified carbonic acids with $R > C_{21}$;

EMKO — carboxy reagent (see below);

NA — naphthenic acids. Liquid NA from kerosene fractions or vat residue of NK ("acidic tar") are used;

DAH — 1,2-diacylhydrazine;

HCA — hydrazine of carbonic acids.

DAH and HCA — chelate-forming collectors. Chelates — neutral, low-soluble in water covalent intra-complex compounds which are formed when organic reagents having one anionic group (-OH, -SH and others) and, at least, one atom-donor of basic character (= N-, = S, = O), articulate metal ions.

It's known that non-ferrous metallurgy and galvanic productions are dangerous sources of environment pollution by non-ferrous metals (NFM). Analysis of the state of purification facilities, accomplished by Radushev V.A. and Chernovaya G.V. (Technical Chemistry Institute, Russian Academy of Sciences, the Urals division, Perm city) selectively on enterprises of Perm city and it region, has shown that major polluting metals (by their quantity descending) are Fe(III) > Cr(III, VI) > Cu > Zn > Ni. Purification facilities, acting mainly by principle of metal sum sedimentation with the help of lime (less often — NaOH) with mud sanitation into clump, are only on part of enterprises. Electro- and galvanic co-agulants, existing on some plants, do not provide waste deep purification till MPC and result in unused wastes. Seemingly evident decision — waste local collection and their processing with the obtaining of being exploited ingredients — cannot be realized practically on existing at the moment industries.

Number of works, devoted to utilization of foam products of ionic flotation, is rather limited. At the same time, qualified application of foam product could decrease rate cost on waste cleaning and even make the scheme profitable. It's known that metallic soaps of carbonic acids have multiple applications — as paintwork components, greases and so on. As a result of two-stage purification according to proposed approach, two kinds of wastes are obtained:

- metal hydroxides of common formula $Me(OH)_2$;
- foam product of simplified formula $nMe(OH)_2 \cdot Me(OH)(RCOO)$, (where $n = 4-16$, and $RCOO^-$ — anion of carbonic acids, contained in flotoreagent EMKO).

The scheme implies the transfer of these compounds into medium soaps $(RCOO)_2Me$ which are highly-soluble in mixture toluene-nefras (1:1).

Process of soap obtaining proceeds in accordance with equations 8.11 and 8.12:

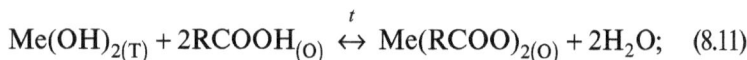

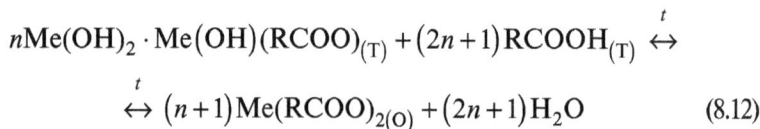

$$Me(OH)_{2(T)} + 2RCOOH_{(O)} \overset{t}{\leftrightarrow} Me(RCOO)_{2(O)} + 2H_2O; \quad (8.11)$$

$$nMe(OH)_2 \cdot Me(OH)(RCOO)_{(T)} + (2n+1)RCOOH_{(T)} \overset{t}{\leftrightarrow}$$

$$\overset{t}{\leftrightarrow} (n+1)Me(RCOO)_{2(O)} + (2n+1)H_2O \quad (8.12)$$

It's necessary preliminary to dry sediments of hydroxides and a foam product. If damp sediments are used, one can't separate water from organic layer (stable water-oily emulsion is being formed). Obtained organic 10–15% solute of NFM medium soaps is used as a marketable product, for instance, in compositions for wood- and metal-coatings. Presence of Cu (II), Zn compounds gives antiseptic properties to the coatings.

Principal scheme of wastewater purification is shown on Fig. 8.20 and represents itself three mutually connected blocks of — NFM sedimentation, polishing of filtrates by pressure-head flotation, obtaining of NFM metal soap solute.

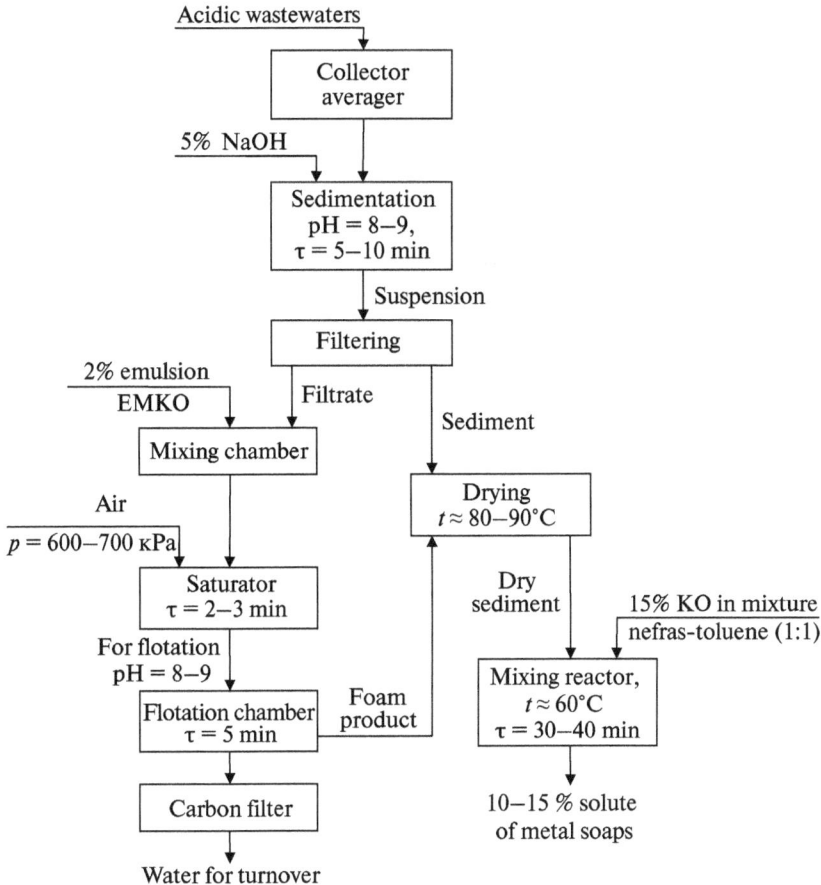

Fig. 8.20. Principal technological scheme of industrial effluents
from NFM with utilization of sediments.
*KO — abbreviation for special reagent

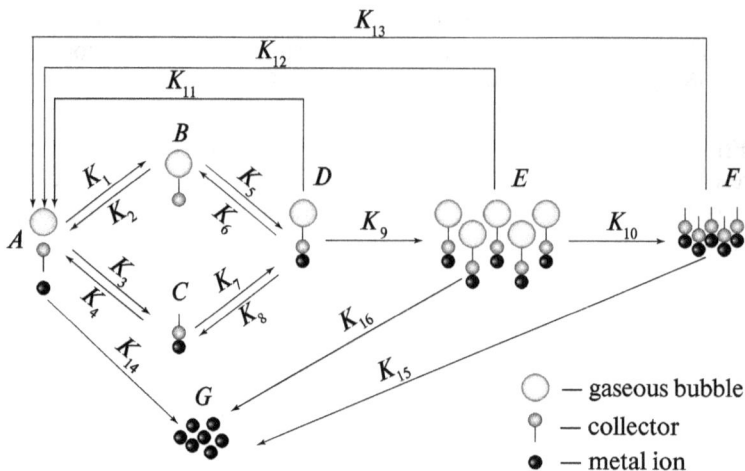

Fig. 8.21. Scheme of process of metal extraction by ionic flotation by Ksenofontov B.S. [31]. *A, B, C, D, E, F, G* — process stages; K_1–K_{16} — constants, characterizing speeds of transfers between flotation process stages

Ions Cr (IV) must be reduced before sedimentation, for instance, with ammonium hydrosulfate:

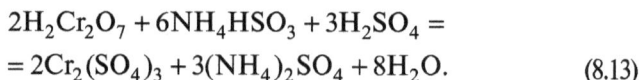

$$2H_2Cr_2O_7 + 6NH_4HSO_3 + 3H_2SO_4 =$$
$$= 2Cr_2(SO_4)_3 + 3(NH_4)_2SO_4 + 8H_2O. \qquad (8.13)$$

Known from literature scheme of wastewater purification from NFM includes sedimentation of major quantities of metals with alkali, polishing of filtrates with pressure-head flotation with carboxy collector EMKO, transfer of obtained residues into medium soaps of carbonic acids Me(RCOO)$_3$ in the mixture of organic solvents, used like paintwork component as well as in compositions for protective wood- and metal-coverings. Meanwhile, it's been shown that at realization, the scheme can be self-repaid.

Major drawback of Golman A.M. model, as demonstrated above, is absence of considering of flotocomplex formation as the process individual stage. But namely flotocomplex plays major role in flotation process.

Let consider in more details the flotation multistage model taking in account major specificities of the process, including also flotocomplex formation and process proceeding in flotodefecating regime. Flotation process major stages according to this model are shown on Fig. 8.21.

Equation system, describing above-pointed process, has the following form (8.14):

$$\frac{dC_A}{dt} = -K_1 C_A + K_2 C_B - K_3 C_A + K_4 C_C + K_{11} C_D + K_{12} C_E + K_{13} C_F - K_{14} C_A$$

$$\frac{dC_B}{dt} = K_1 C_A - K_2 C_B - K_5 C_B + K_6 C_D$$

$$\frac{dC_C}{dt} = K_3 C_A - K_4 C_C - K_7 C_C + K_8 C_D$$

$$\frac{dC_D}{dt} = K_5 C_B - K_6 C_D + K_7 C_C - K_8 C_D - K_9 C_D - K_{11} C_D \qquad (8.14)$$

$$\frac{dC_E}{dt} = K_9 C_D - K_{10} C_E - K_{12} C_E - K_{16} C_E$$

$$\frac{dC_F}{dt} = K_{10} C_E - K_{13} C_F - K_{15} C_F$$

$$\frac{dC_G}{dt} = K_{14} C_A + K_{15} C_F + K_{16} C_E$$

Proposed system must satisfy at least two conditions, namely, at the initial time moment, colligand concentration on the first stage must be equal to initial concentration in solution, and sum of colligand concentrations by all stages must be equal to its initial concentration at any time moment.

Such system, as a rule, is settled with the use of numerical methods.

Efficiency of extraction of particular metals from wastewaters, obtained by numerical method and defined experimentally, is presented in Table 8.8. Comparison of calculated and experimental values shows little discrepancy not exceeding 7% that allows to exploit numerical data for evaluation of efficiency of wastewater purification, including that from metals, with the use of ionic flotation.

It's worth to note that this possibility allows to define being extracted substance concentration on each of the process considered stages at any time moment without pursuing high-cost experiments that is specifically important at projecting wastewater purification system. Besides, there is a possibility to find stage limiting by time, influencing which flow, the common time of pollution extraction process, including one of metal ions, can be pruned down.

Table 8.8. Indicators of extraction of individual metals from waste-waters with the use of ionic flotation

N	Metal	Metal concentration in wastewater, mg/l	Flotation time, min	Purification efficiency determined experimentally, %
1	Chrome (common)	2.2	15.5	91.4
2	Lead	4.4	15.5	89.6
3	Nickel	3.5	15.5	92.7
4	Wolfram	2.9	15.5	93.3
5	Cobalt	4.1	15.5	89.6

One of the ionic flotation advantages is a selectivity high degree. At that, the selectivity can be increased, regulating media conditions (pH, temperature). The selectivity can be connected also with the specificities of collector molecular composition and its concentrations in a solute. Thus, at collector addition in very small doses, ions of one sign and charge can be divided.

Other advantage of ionic flotation from the point of view of waste purification is that at extraction from water, for example, of heavy metals, surfactants, presenting in effluents: oils, soaps etc, are extracted simultaneously.

Ionic flotation important property is its capability of concentrating particles of colloidal size from extremely diluted solutes. Ions are easily extracted at concentration from shares to hundreds of milligrams per liter. Together, one should remember that from economic point of view, quantity of a collector, taken for ionic flotation, is proportional to being extracted element quantity (and depends on its state) but doesn't depend on solute quantity.

Besides, needed quantity of collector might be rather small at correct selection of the process holding conditions (pH, in particular). For instance, it's necessary just 0.01 mole of anionic collector for utter extraction of 1 mole of cuprum at pH = 6 according to the research results.

One of the ionic flotation advantages is that it doesn't demand machines with moving parts. This circumstance implies potential possibility of its usage for the purification of nuclear electro-stations' effluents. The matter is that the machines with moving parts, filter-presses or riddles require constant deactivation that is uncomfortable. Besides, the ionic flotation is well incorporated into treatment of solutes containing radioactive materials because all of the process can be held in closed containers (just from time to time it's necessary to collect scum and then to process it). In this regard, significant number of investigations is devoted to radioactive elements removal (uranium, radium, isotopes of strontium, cesium, rare-earth metals and so on) from wastewaters of scientific research laboratories and industrial enterprises.

Finally, the process doesn't demand solute big quantities. Instead of this, only various apparatuses for air bubbles transmission are needed which can be fixed in regular gutterways.

Thus, wastewater purification from heavy metals with the use of ionic flotation, from one side, allows to delete them efficiently from water by simple and by relatively not expensive operations, and from the other side, to extract the most precious elements for further usage.

Comparison with other methods of wastewater purification from heavy metals shows that ionic flotation combines advantages of extraction and ionic exchange. Ionic flotation according to kinetic characteristics is on a liquid extraction level and is close to ionic exchange by the amount of losses of organic substance. Meanwhile, ionic flotation possesses high productivity. In comparison with coagulation and sedimentation methods, the ionic flotation is 4–6 times more productive, demands less spaces, is the universal method for removal of suspended particles.

Nevertheless, ionic flotation hasn't got such wide spread at wastewater purification as, for instance, reagent methods or an extraction. It's connected, first of all, with the earlier discussed complexity of collector choice as well as with difficulties of regeneration of reagents. Suggested algorithm permits to formalize and speed up collector selection process.

One of the serious difficulties on the way of industrial introduction of ionic flotation — is relatively high cost of some surfactants [31]. In this regard, the necessity to regenerate surfactants appears. Nevertheless, extracted metal cost sometimes is so high that expenditures for surfactants can be ignored. Together, as was aforesaid, collector necessary amount can be lessened at literate selection of process parameters.

Let consider improvement ways and the perspectives of ionic flotation usage in the practice of wastewater cleaning from heavy metal ions.

For to develop ionic flotation as the way of wastewater purification from heavy metal ions, it seems expedient the development and improvement of the methodology of optimal collector choice. Meanwhile, it should meet all requirements of the process as well as be cheap, available and regenerable under a need.

Correct choice of the process flow conditions often allows to increase component extraction efficiency as well as to cut collector rate. Hence, it's necessary to pay special attention to theoretical developments in this sphere.

Furthermore, as was afore-underlined, for final choice of both, a collector or process flow optimal conditions, it's needed to pursue experiments. It means that the improvement of experimental installations and laboratory research holding schemes will allow to get fuller evidences on the process and laws which should be accounted for at its pursuing.

Development of chamber designs to hold the ionic flotation in industrial scales represents as well one of the important tasks for practical application of the ionic flotation.

Various ways of processing and usage of ionic flotation wastes, for instance, for to obtain metallic soaps, deserves attention as well because, meanwhile, from one side, amount of wastes decreases, but from the other side, marketable product is obtained allowing to make in some cases the cleaning scheme to be self-repaid.

At last, combination of purification various methods permits to reach the best results rather often. Thus, junction of ionic flotation with reagent method or, for instance, with ion exchange is the most promising.

Undoubtfully, ionic flotation methods are promising especially at extraction of non-ferrous, rare and noble metal small quantities and deserve bigger attention from researchers' and technologists' side [31].

9 EXPERIENCE AND PERSPECTIVES OF FLOTATION COLUMNS USAGE IN ENRICHMENT OF MINERALS AND WASTEWATER PURIFICATION PROCESSES AT APPLICATION OF WATER-USE RECURSIVE SYSTEMS

Recommendations, deriving from flotation multistage model, were used at creation and testing of flotation column for purification of waste and recursive waters from fine hydrophobic pollutions (Fig. 9.1). Deep jet aerators of vertical kind, working by principle of ejection of atmospheric air by jets of being purified air, were used in the developed column. The column upper part is narrowed for foam product density increase and washing water rate decrease. Deletion of the possibility of entrainment of fine mineralized bubbles with chamber product is provided therein by installation of lamellar catchers of bubbles which usage ensues directly from flotation multistage model. To pursue trials, test-industrial sample of flotation column with working volume of 1 m³ was made.

The column works in the following way. Being purified pulp is given via nozzle under pressure of about 0.2 MPa into vertical tubes *1*, having holes in their upper part for atmospheric air pass. Jets, being formed by caps, eject air from atmosphere and move down together with it till their exit into bottom part of flotation chamber *2*. At joined movement of pulp and air, the air dispersion on tiny bubbles proceeds with their simultaneous mineralization. Aerated flows in bottom part of chamber *2* are transformed into lifting aerated flows wherein bubble mineralization process continues. Mineralized bubbles in chamber upper part surface-emerge into foam layer, being formed inside cylinder *3* wherefrom they are deleted through foam sill into gutter *4*. Chamber product through ledges of flotation chamber *2* enters zone of fin catchers *5* of bubbles, which movement through, mineralized bubbles, being late to surface-emerge, lift up underneath upper, inclined towards foam collector *3*, fins and are directed into foam layer. Purified liquid is directed along side compartments *6* into bottom part is and unloaded into columns via unloading device *7*.

Adopted in the developed column the way of bubbles mineralization in jet aerators and in lifting-up aerated flows in work chamber of the column, the installation of laminar catcher of bubbles provides for particle effective flotation at column smaller height at high speeds of movements of pulp or being purified liquid.

Scheme of semi-industrial sample of flotation column FK-1 is presented on Fig. 9.1.

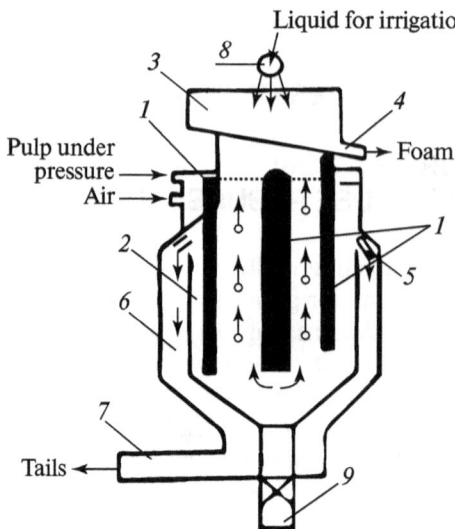

Fig. 9.1. Scheme of semi-industrial sample of flotation column FK-1:
1 – deep tubularjet aerators;
2 – flotation camera; *3* – foam collector; *4* – foam chute; *5* – platelike bubble catchers; *6* – unloading compartments;
7 – unloading sleeve; *8* – irrigator;
9 – emergency tap

Trials of laboratory and semi-industrial devices were held on the sink of thickener, working on fugate of centrifuges, dehydrating end concentrate of flotation (KCl). Thickener sink contained mainly fine hydrophobic particles of 0.1 mm size. Results of trials are shown in Table 9.1.

Trials have established that, as a result of flotation purification of thickener sink, 94.2–98.7% of KCl is extracted into foam with content of 93.6–97.43% of KCl therein. The obtained foam product is close by quality to issued industrial concentrate (95% KCl). That is why it can mix with ready concentrate, providing increase in KCl extraction. Chamber product, presenting "queen cell", cleaned from pollutions, can be directed into process head and other points of sylvite ore enrichment scheme. Received spec. productivity (18

Table 9.1. Results of trials of flotation column FK-1

Product	L:S (liquid: solid)	Content, %		Extraction KCl, %	Efficiency, m³/h
		KCl	H₂O		
Results of flotation on laboratory installation					
Thickener sink	18.5	86.5	Not defined	100	
Flotation foam product	1.63	93.7	2.7	98.6	0.7
Chamber product	604	10.4	0.13	1.4	
Results of flotation on semi-industrial installation					
Thickener sink	19.4	96.89	1.39	100	
Flotation foam product	8.8	97.36	1.28	94.1	9.6 (18 m³/m²h)
Chamber product	157.8	12.4	6.18	5.9	

m³/m²h) of trialed column exceeds spec. productivity of standard pressure-head flotation skimmers wherein it doesn't exceed 6–8 m³/m²h because of low speed of microbubbles surface emerging.

Recommendations have been worked out on the basis of trials results for industrial flotation column creation for ore pulps flotation and cleaning of waste and recursive waters in various regimes with aeration by atmospheric air or some other gas.

One should underline especially that water recursive cycles are one of the major priorities of water rational use. Meanwhile, much significance is also in cost of such task settlement. In some cases, when strict necessity of water-use recursive system introduction is under demand, economic matters become less important. Nevertheless, economic questions become important and defining for majority of cases of practical introduction of recursive systems.

In this regard, choice of technologies, allowing to reach set technical requirements at minimal costs, has been held. Such technologies, as a rule, are called the best available technologies.

Technological scheme of cleaning of wastewaters of paintwork industries has been developed on the basis of made research on the efficiency of applied methods. Such scheme includes reagent treatment by the way of input of 3 reagents (K, AK, FP) in definite doses, next defecating of reagent treated wastewater in dirt-oil-catcher of defecating type, afterdefecating of water in flotation skimmer with water afterpurification (polishing) in filters with grainy and carbon loads. Hereat, mixing with reagents, defecating and flotation proceed in sole flotoharvester, wherethrough being purified drain goes of its own accord.

Equipment for the purification of wastewaters of paintwork industry according to afore-considered technological scheme (Fig. 9.2) is presented vividly on the photo (Fig. 9.3 a — 9.3 b).

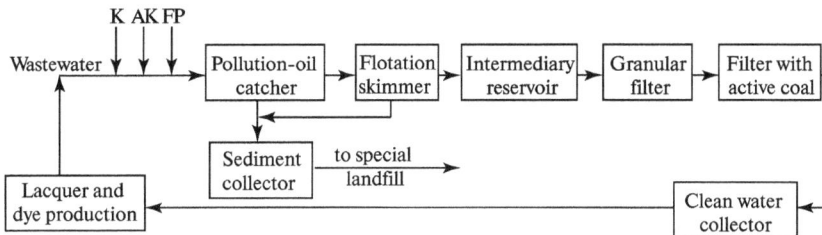

Fig. 9.2. Principled scheme of purification of recursive water
of paintwork production

Fig. 9.3. *a* — Common view of flotoharvester for purification
of wastewaters of paintwork production (front view);
b — common view of flotoharvester for purification
of wastewaters of paintwork production (from-above view)

Efficiency of purification with the use of above described technological scheme is shown in Table 9.2.

Table 9.2. Efficiency of purification
of wastewaters of paintwork production

Indicators	Water quality, mg/l		Requirements to recursive water, mg/l
	Initial	Purified	
Nitrogen of ammonium	19.3	2.48	12.3±1.3
Ammonium-ion	24.9	3,2	15.8±1.6
Suspended substances	9250	<5	130±13
Fats	5.4	2.1	5.3±0.5
Mineral oils	92	<0.05	1.2±0.12
Common chrome	0.066	<0.002	0.0084±0.0009

Data demonstrated in Table 9.2 testify on the achievement of recursive water needed quality with the possibility of its further usage in paintwork production.

These results as well point out the possibility of usage of water of similar quality in other technological processes of construction industry and other industrial branches. Big interest, in this regard, is in wastewater cleaning of consumer technique production.

Wastewater, containing also surfactants, comes into accepting reservoir (stance *1*, look Fig. 9.4). When water level in reservoir reaches maximum, swimming pump (stance *H*1) is switched on automatically and pumps wastewater into decanter which is the simplest kind of flotoharvesters (stance *3*).

Reagents for wastewater cleaning are supplied on flotoharvester entrance chamber: by pump-dispenser 211 — lime 5% solution; by pump-dispenser 221 — coagulant 5% solution (aqua-aurate 30); by pump-dispenser 231 — flocculant 0.05% solution (praestol 655). Entrance chamber is a conditioning chamber and serves for intensive mixing and rational injection of reagent into being purified water.

As a result of reagents impact on pollutions available in wastewater, there proceeds formation of flakes and their further separation from water in flotation and sedimentation zones which in, working space of flotoharvester is divided. At the expense of air bubbles contacting with the flakes, possessing hydrophobic properties, formation of flotocomplexes takes place with their further surface-emerging into foam layer.

Working liquid is supplied into flotation zone with the help of special aeration pump *H*3 which takes water from intermediate reservoir, stance *4*. Meanwhile, the air supply is going on into pumping main of pump, stance *H*3, at that, the air quantity shouldn't exceed of about 5% of water-air mixture rate. Sedimentation of particles and flakes not caught by flotation takes place in defecating zone.

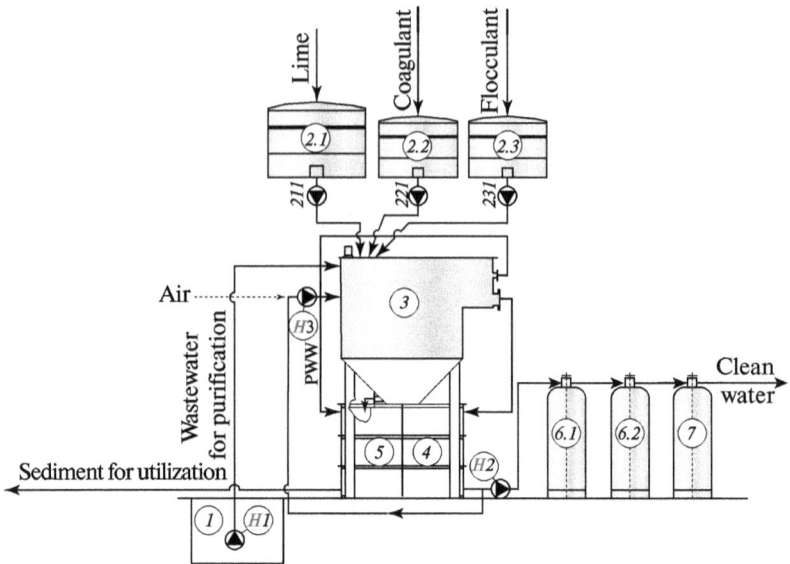

Fig. 9.4. Scheme of purification of wastewaters with flotoharvester application , PWW — purified wastewater

Caught sediment and flotation scum, which are directed periodically into scum-storage, stance *5*, wherefrom are directed on utilization, are formed in flotoharvester as a result of cleaning of wastewaters.

After purification of wastewaters in the flotoharvester, water flow is supplied of its own accord into intermediate reservoir, stance *4*, equipped with automatic pump *H2* which is turned on as intermediate reservoir fills in and supplies the water on mechanical, stance *6.1*, *6.2*, and sorption filters, stance *7*. Water polishing from suspended substances and dissolved impurities proceeds in the filters. Purified water is thrown out to canalization after the filters.

The offered technological scheme uses the flotoharvester which scheme is shown on Fig. 9.5. The distinctive feature of this flotoharvester is the device in flotation and defecating zone of fine-zoned blocks with various inter-shelf distance (Fig. 9.5, stance *4*, *6*). Usage of such blocks allows to optimize hydrodynamical regime in the flotation and defecating zones.

Lime preparation deserves a special attention. Experimental research held by us has shown that it's necessary to use apparatus with blender for better preparation of fine-dispersed suspension of lime, at that, with compressed air supply into mixer work zone in some cases (Fig. 9.6*a* and 9.6*b*).

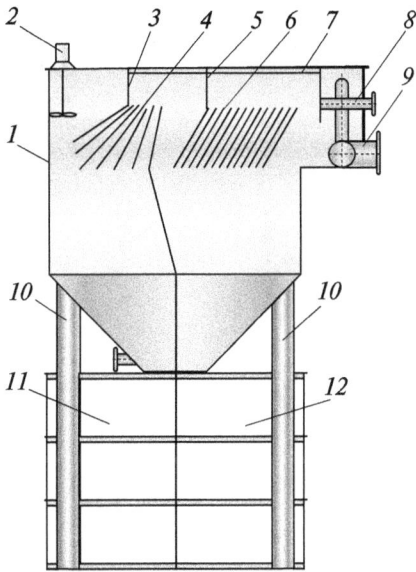

Fig. 9.5. Scheme of flotodecanter:
1 — shell; *2* — mechanical blender;
3,5 — watersemi-burried walls;
4 — shelves of fine-zoned defecating
of flotation zone; *6* — shelves of
fine-zoned defecating of defecating
zone; *7* — foam sill; *8* — sleeve of
foam product bend; *9* — sleeve of
purified water bend; *10* — supports;
11 — intermediate reservoir for purified water collection; *12* — reservoir
for foam product collection

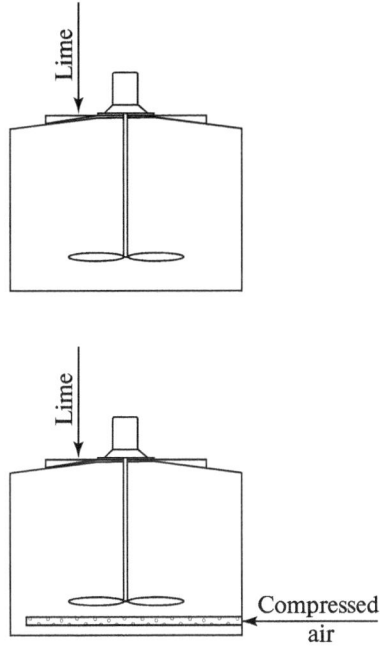

Fig. 9.6. *a* — apparatus for lime
preparation with blender;
b — apparatus for lime
preparation with blender and
compressed water supply

Compressed water supply into blender rotation zone allows to improve suspension of lime fine-dispersed particles and, hence, to exclude sedimentation of these particles in so-called "dead" zones, for instance, on remote periphery. Used system of reagent preparation and mixing, first of all, of lime, permits to use reagents rationally, reaching thus the maximal effect of their usage.

Held industrial trials of flotodecanter sample of such type have shown efficiency while usage in wastewater purification processes of consumer technique production (Fig. 9.7–9.9).

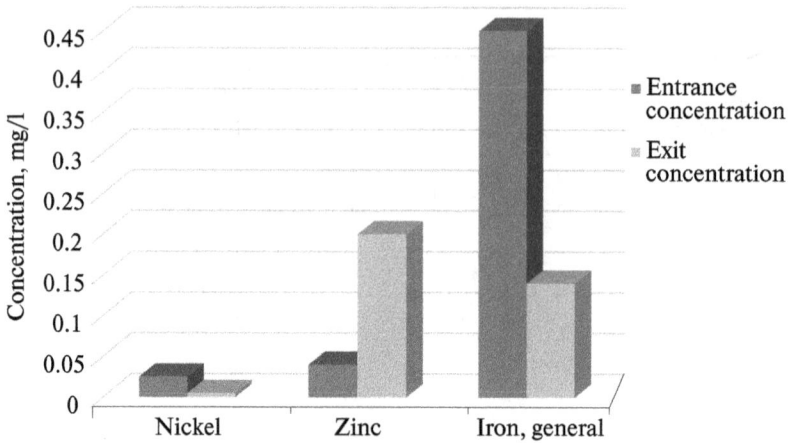

Fig. 9.7. Extraction of metals

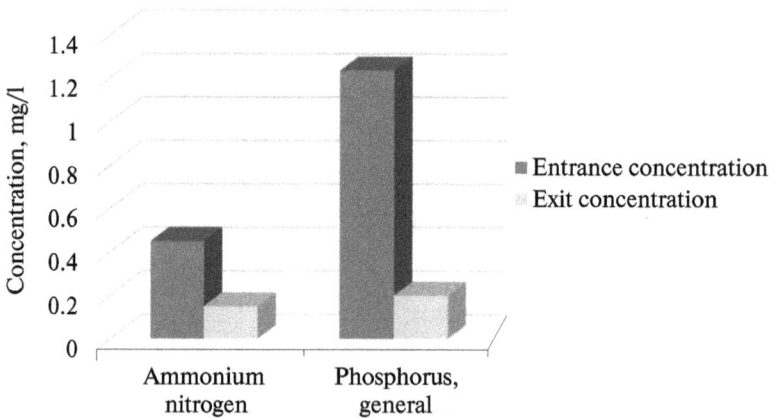

Fig. 9.8. Extraction of metals of ammonium nitrogen and phosphorus

Presented data on Fig. 9.7–9.9 testify high efficiency of wastewater purification with the use of flotoharvesters. The process intensification in event of necessity can be reached by the way of usage of additional knots, for instance, by installation of electro-processing blocks and so on. In some cases, efficiency of purification because of the usage of noted additional blocks might be raised on 40–50 %.

It's worth to note specifically that held by us research on wastewater cleaning with the use of flotation technologies has shown that the reagent flotation

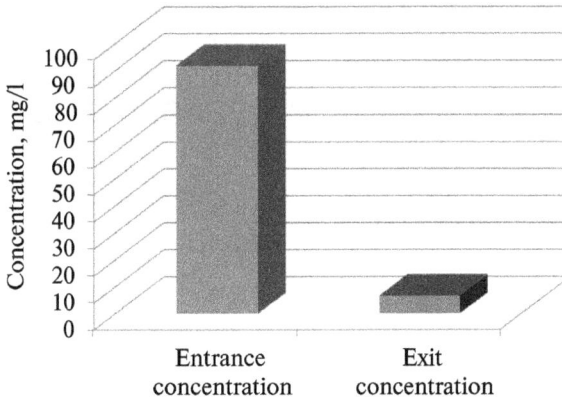

Fig. 9.9. Extraction of suspended substances

is feasibly to combine not only with defecating and filtration but also with different kinds of physical impacts, for instance, with the use of electrical and magnetic fields, ultraviolet, ultrasound and so on. Theoretical analysis and experimental data demonstrate that, in this case, the maximal impact can be reached at minimal energy expenditures. For to use such complex technologies we are working on new types of flotoharvesters.

10. INTENSIFICATION OF WASTEWATER FLOTATION PURIFICATION IN VARIOUS PRODUCTIONS

10.1 INTENSIFICATION OF WASTEWATER FLOTATION PURIFICATION PROCESSES OF SOAP-BOILING PRODUCTION

Questions of environment safety have large significance for functioning of soap-boiling productions, including questions of wastewater purification. Waste, which is necessary to be purified before dropping into canalization, is formed in the process of soap production.

Obtaining of soap on some domestic soap-boiling productions is pursued in periodical regime by the way of load into boiling cauldron of necessary ingredients. Meanwhile, major part of wastes, created on such productions, is supplied from soap-boiling workshop (60%). 30% of drainages come from fat decanters of workshop n.3 and about 10% — from workshop n.1. Approximate principled scheme of waste collection is shown on Fig. 10.1.

Purification of wastewaters is made by the way of their supply into fat-catcher representing sectioned decanter. Fat-catcher working principle is in the following: surface-emerging of fat-soluble fraction (fat, soap) proceeds while wastewater defecating. As accumulations of this fraction in fat-catcher upper part proceed, the fraction deletion by means of sucking with the help of vacuum and its supply into boiling cauldron (workshop n.2) goes on. Mass temperature in fat-catcher is maintained in 60–70 °C limits. Liquid, cleared in fat-catcher, then is being dropped into a well wherefrom it's directed into a city collector.

Chemical analysis of wastewaters of soap-boiling plant before and after their purification was held basing on the series of indicators, including definition of COD (Chemical Oxygen Demand), suspended substances, fat and soap.

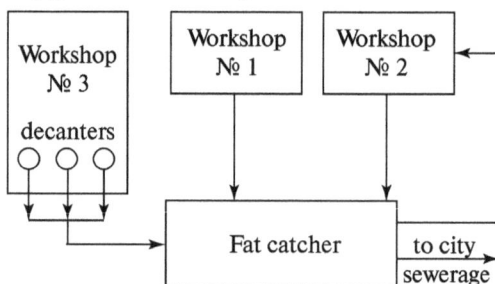

Fig. 10.1. Production principled scheme of soap-boiling industry: workshop n.1 — production; workshop n.2 — boiling cauldron; workshop n.3 — decanters

Table 10.1. Content of initial wastewaters of soap–boiling production

Sample n.	Place of sample choice	Ingredients			
		Fat, mg/l	Soap, g/l	COD, mgO$_2$/l	Suspended substances, g/l
1	fat catcher	714	40.3	27264	49.3
2	fat catcher	572	36.9	27100	52.0
3	fat catcher	400	136	32000	62.0

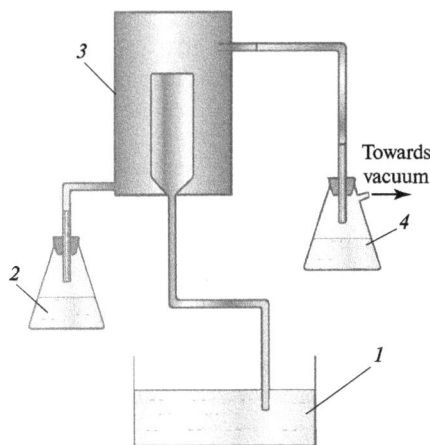

Fig. 10.2. Laboratory installation scheme for vacuum flotation pursuing:
1 — capacity for initial drainages;
2 — capacity for purified liquid;
3 — vacuum apparatus; 4 — capacity for foam product

Wastewater content was being determined with the help of methodologies applied in laboratory practice for the analysis of industrial wastewaters.

As a result of initial wastewater analysis, we have got the following data (Table 10.1).

As data of chemical analysis show, wastewaters of the soap-boiling production have high degree of the pollution. This is mainly impurity with organic substances, largely — with soaps. By exterior, unpurified drainage sample represents viscous and soapy to the touch liquid of brownish color. It thickens till pastelike state at the cooling till room temperature.

Investigations on possibility of wastewater polishing was being held by ways which are used the most frequently. In particular, the following indicators can be referred to flotation advantages: continuity of the process, application wide scope, not high capital and exploitation expenditures, apparatus formalization simplicity, large speed of the process in comparison with the defecating. In lab conditions, we were holding experiments on polluted wastewater purification by the ways of vacuum and mechanical flotation.

Fig. 10.2 shows laboratory installation scheme for the vacuum flotation. Method essence, as it's known, is in the creation of oversaturated solution of air in the liquid. At pressure reduction, air bubbles are extracted from the solution, the bubbles float the pollutions.

Fig. 10.3 shows lab installation scheme for the flotation with air mechanical dispersion. Mechanical dispersion of the air in the flotation machines in the case of fat-containing wastewater purification provides not only for gas bubble uniform spread across volume but also for factually utter absence of "dead zones" that is rather important in this case.

Last time, such installations have been used for the purification of wastewaters with high content of suspended particles and of fat.

Besides, experiments with preliminary filtration of initial drainages and subsequent flotation of filtrate with fresh water addition were held. Drainages, cleaned with various methods, were analyzed by us mainly on fat and soap content. The laboratory research results are shown in Table 10.2.

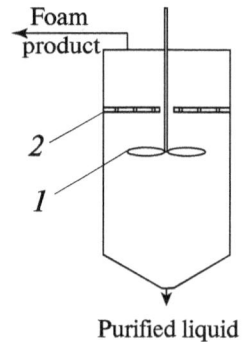

Fig. 10.3. Laboratory installation scheme for flotation with air mechanical dispersal: *1* — impeller; *2* — ebullient layer lattice

Table 10.2. Content of wastewaters of Moscow Soap–boiling Plant after purification on flotation laboratory installations

Product	Results of chemical analyses			Conditions of experiment holding
	Fat, mg/l	Soap, mg/l	Suspended substances, g/l	
Initial drainages	714.0	40300.0	49.3	Vacuum
Purified liquid	152.0	14100.0	1.402	flotation
Initial drainages	106.4	1060.0	2.401	Mechanical
Purified liquid	6.1	628.3	0.24	flotation
Initial drainages	53.2	530.0	−	Mechanical
Purified liquid	11.2	294.8	−	flotation
Initial drainages	106.4	1060.0	2.401	Vacuum
Purified liquid	8.2	735.4	0.36	flotation
Initial drainages	53.2	530.0	−	Vacuum
Purified liquid	10.3	271.8	−	flotation
Initial drainages	129.5	959.0	−	Mechanical
Purified liquid	12.6	400.4	−	flotation
Initial drainages	129.5	950.0	−	Vacuum
Purified liquid	12.6	180.8	−	flotation

Pursued investigations on the analysis of initial wastewaters of Moscow Soap-boiling Plant have shown that high values of the series of ingredients, for instance, concentrations of soap, fats, suspended substances, organic compounds, evaluated by COD, are observed. To normalize the content of dropped wastewaters which in the row of cases, represents concentrated aqua colloidal systems, it's necessary to hold organizational technological measures, to improve fat-catcher exploitation thus but to pursue polishing with the help of pressure-head flotation apparatus.

The pursued investigations in the laboratory conditions with the use of various flotation apparatuses have demonstrated that at flotation polishing, fall of content of fats by 10 and more times and of soap by 4–5 times takes place.

Thus, held investigations have confirmed flotation usage feasibility on the stage of wastewater polishing.

10.2 INTENSIFICATION OF FATTY WASTEWATER FLOTATION PURIFICATION WITH THE USE OF REAGENTS

Investigations pursued by us have shown that flotation purification of fat-containing wastewaters can be essentially intensified by the usage of reagents.

Pursuing of research in laboratory conditions was made by the way of choice of wastewater samples after workshop fat-catchers of Plant Oil Factory and before Local Purifying Facilities (LPF) of this factory. Samples of drainages after workshop fat-catchers were being chosen with the purpose of purification of these wastes according to a model of fat-catcher with fine-zoned defecating (clearing) block. Nevertheless, lab research with these samples didn't give purification noticeable effect because drainages, going out from the fat-catchers, do not respond to further cleaning without reagents. In this regard, major work was held with mixed drainages before LPF.

Held research methodology was in the accomplishment of further sequence of operations. Wastewater samples were a subject of reagent treatment. Salts of aluminum (vitriolic aluminum, aluminum hydroxy-chloride) and of iron (chlorous and vitriolic iron) as well as flocculants — Praestol (production of German Federal Republic and of Russian Federation, Perm city) and Zetag (GFR production) were qua reagents. It was established as a result of preliminary tests that delineated reagents, excluding aluminum salts, did not give purification noticeable effect. Incidentally, all further investigations were pursued with vitriolic aluminum as factually at the same water clearing effect, the vit-

Table 10.3. Results of wastewater purification by defecating at addition of vitriolic aluminum $Al_2(SO_4)_3 \cdot 18H_2O$ (drainage choice before LPF; water purification by defecating in lab cylinders with working volume of 50 ml during 1 hour)

N.	Dose $Al_2(SO_4)_3 \cdot 18H_2O$, g/l	Concentration of pollutions in wastewater, mg/l			
		Fats		Suspended substances	
		Initial water	Purified water	Initial water	Purified water
1	0.5	135.0	133.2	680.7	676.3
	0.7	135.0	132.9	680.7	653.4
	0.9	135.0	41.2	680.7	183.5
	1.0	135.0	39.6	680.7	112.7
	1.1	135.0	33.4	380.7	98.4
	1.2	135.0	30.8	680.7	88.5
2	0.9	183.9	48.7	656.3	110.8
	1.0	183.9	43.4	656.3	101.3
	1.1	183.9	41.2	656.3	90.4
3	0.9	147.3	38.1	539.1	108.7
	1.0	147.3	36.7	539.1	97.6
	1.1	147.3	33.5	539.1	89.3
4	0.9	191.4	57.6	710.5	238.9
	1.0	191.4	49.2	710.5	208.7
	1.1	191.4	42.7	710.5	197.5

riolic aluminum is cheaper that aluminum hydroxy-chloride by 20–25 times approximately. After the treatment with the vitriolic aluminum, the wastewater was being cleared consistently by defecating in cylinders of working volume of 500 ml during 1 hour. Then the cleared water was being floated in a lab flotation machine during 30 minutes. The cleared water after flotation was then being polished on filters with combined fill from gravel and carbon in 1:1 ratio at filtration speed of 10 m/h.

The cleared water analysis after each step of purification was held with the use of standard (typical) methods affirmed in RF. The results of pursued investigations are shown in Tables 10.3–10.6. Meanwhile, Table 10.3 demonstrates data concerning the purification of wastewaters from the specific pollutions — fats and suspended substances — with the use of purification one step — a defecating. It's seen from the data shown in this table that noticeable effect of cleaning starts to show up at vitriolic aluminum dose of 0.9–1.2 g/l

Table 10.4. Efficiency of wastewater purification from fats by purification steps defecating–flotation–filtering (drainage choice before LPF; addition of vitriolic aluminum $Al_2(SO_4)_3 \cdot 18H_2O$ into drainages before defecating)

N.	Dose $Al_2(SO_4)_3 \cdot 18H_2O$, g/l	Concentration of pollutions in wastewater, mg/l			
		Initial water	After defecating	After flotation	After filtering
1	0.9	135.0	42.1	33.8	19.2
	1.0	135.0	39.6	28.4	12.8
	1.1	135.0	33.4	25.7	10.3
2	0.9	183.9	48.7	40.5	9.8
	1.0	183.9	43.4	38.8	8.6
	1.1	183.9	41.2	32.6	8.5
3	0.9	147.3	38.1	29.2	10.4
	1.0	147.3	36.7	24.7	9.5
	1.1	147.3	33.5	19.3	9.4
4	0.9	191.4	57.6	48.2	10.9
	1.0	191.4	49.2	33.8	9.7
	1.1	191.4	42.7	28.5	8.9

calculated per $Al_2(SO_4)_3 \cdot 18H_2O$. Vitriolic aluminum doses $Al_2(SO_4)_3 \cdot 18H_2O$ in interval of 0.9–1.1 g/l give stable effect at wastewater clearing from fats by scheme defecating-flotation-filtering (Table 10.4). Also, stable data are observed at definition of concentrations of fats and suspended compounds after purification concluding stage — a filtering (Table 10.5).

It's known that common task is the purification not only from fat and suspended substances but also from dissolved organic and mineral compounds. Combined data, testifying efficiency of purification with the use of vitriolic aluminum are shown in Table 10.6.

Pursued experimental investigations have shown that wastewater purification stable effect is observed at coagulant doses of 1.0–1.1 kg per 1 m³ of drainages. At these doses, values of major indicators of wastewater purification quality do not exceed values of action levels of pollution major substances which by, dropping of purified wastewaters into city canalization of Valuyky town is controlled.

In the case of coagulant addition in less doses (less than 1.0 kg per 1 m³ of drainages), unstable effect of cleaning by major indicators, including by number of fat fractions, is observed.

Table 10.5. Impact of dose of vitriolic aluminum $Al_2(SO_4)_3 \cdot 18H_2O$ on content of fat and suspended substances in wastewater (drainage choice before LPF; water purification by scheme defecating–flotation–filtering)

N.	Dose $Al_2(SO_4)_3 \cdot 18H_2O$, g/l	Concentration of pollutions in wastewater, mg/l			
		Fats		Suspended substances	
		Initial water	Purified water	Initial water	Purified water
1	0.5	135.0	124.3	680.7	643.5
	0.7	135.0	118.6	680.7	631.2
	0.9	135.0	19.2	680.7	97.1
	1.0	135.0	12.8	680.7	63.5
	1.1	135.0	10.3	680.7	51.9
	1.2	135.0	9.2	680.7	40.8
2	0.9	183.9	9.8	656.3	79.5
	1.0	183.9	8.6	656.3	77.1
	1.1	183.9	8.5	656.3	73.0
3	0.9	147.3	10.4	539.1	108.4
	1.0	147.3	9.5	539.1	93.7
	1.1	147.3	9.4	539.1	93.1
4	0.9	191.4	10.9	710.5	148.3
	1.0	191.4	9.7	710.5	117.2
	1.1	191.4	8.9	710.5	95.8

According to afore-expounded, it's worth to automize the process of co-agulant supply in the row with investing into high-cost devices of instrumentation or to dose coagulant solution with the dose that beforehand guarantees purification demanded effect. The latter in the current situation seems the most rational approach because of coagulant relative cheapness.

Comparative trials of vitriolic aluminum and alumina impact on wastewater purification have shown that alumina rate at cleaning same effect exceeds vitriolic aluminum rate on no more than 10–12 %. In this regard, taking into account better availability and cheapness, it's possible to use alumina, though it forms sediment at the preparation of working solution.

It's recommended to use vitriolic aluminum or alumina qua a coagulant in doses, respectively, 1.0 and 1.12 kg per 1 m^3 of wastewaters.

Experimental research pursued by us has shown that at addition into wastewaters of vitriolic aluminum in quantities of 0.9–1.1 kg per 1 m^3 of wastewaters, pollution flakes are formed which fall out into sediment and part

Table 10.6. Results of wastewater purification by scheme (coagulation of $Al_2(SO_4)_3 \cdot 18H_2O$)–defecating–flotation–filtering

Concentration of pollutions in wastewater, mg/l

N.	Dose $Al_2(SO_4)_3 \cdot 18H_2O$, g/l	Fats Initial water	Fats Purified water	Suspended substances Initial water	Suspended substances Purified water	BOD5* Initial water	BOD5* Purified water	Ammonium nitrogen Initial water	Ammonium nitrogen Purified water	Phosphates Initial water	Phosphates Purified water	Chlorides Initial water	Chlorides Purified water	Sulfates Initial water	Sulfates Purified water	Synthetic surfactants Initial water	Synthetic surfactants Purified water	Dry residue Initial water	Dry residue Purified water	pH Initial water	pH Purified water
1	0.9	183.9	9.8	656.3	79.5	1320	34	8.8	5.3	11.3	3.6	180.4	128.5	49.3	69.5	3.2	2.0	2730	748	8.8	7.0
	1.0	183.9	8.6	656.3	77.1	1320	21	8.8	5.1	11.3	3.0	180.4	128.3	49.3	71.4	3.2	2.0	2730	810	8.8	6.9
	1.1	183.9	8.5	656.3	73.0	1320	16	8.8	5.0	11.3	2.4	180.4	120.9	49.3	75.6	3.2	1.6	2730	836	8.8	6.8
2	0.9	147.3	10.4	539.1	108.4	1270	20	8.2	6.1	19.1	4.3	216.0	139.5	68.9	79.8	4.7	3.7	2648	710	8.0	6.7
	1.0	147.3	9.5	539.1	93.7	1270	20	8.2	5.9	19.1	3.1	216.0	133.6	68.9	87.6	4.7	2.7	2648	745	8.0	6.5
	1.1	147.3	9.4	539.1	93.1	1270	15	8.2	5.9	19.1	2.3	216.0	130.0	68.9	95.3	4.7	2.1	2648	768	8.0	6.5
3	0.9	191.4	10.9	710.5	148.3	1560	38	9.1	8.3	24.5	4.6	98.4	91.4	50.1	68.9	2.9	2.1	2560	697	8.3	6.9
	1.0	191.4	9.7	710.5	117.2	1560	21	9.1	7.8	24.5	2.9	98.4	90.8	50.1	72.3	2.9	1.8	2560	713	8.3	6.8
	1.1	191.4	8.9	710.5	95.8	1560	18	9.1	7.7	24.5	2.2	98.4	90.6	50.1	80.2	2.9	1.8	2560	789	8.3	6.6

*Biological oxygen demand during 5 days.

211

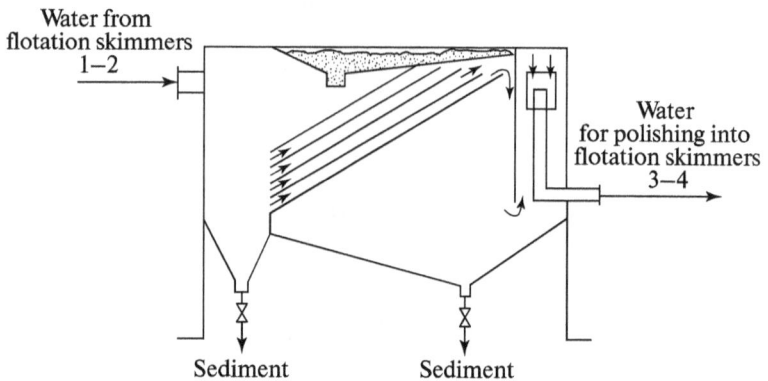

Fig. 10.4. Scheme of modernized fat-catcher

of them is floated. This research results and insight into existing purifying facilities have become the basis for choice and substantiation of technological scheme of wastewater cleaning, proposed for an industrial introduction.

In this scheme, flotodecanter is provided — a fat-catcher of unique design with the use of fine-zoned block (Fig. 10.4) which, as trials have shown, is distinguished by rather high efficiency, regime stability and reliability in exploitation.

Presence of fat in wastewaters jointly with surfactants even more aggravates oxidation conditions of organic compounds by microorganisms. Held by us trials and industrial introduction of flotation technology of wastewater purification on food industry enterprises have demonstrated that it's possible to extract surfactants from the water till the residual concentrations of about 1.5–2.5 mg/l, and fats — till 15–20 mg/l and less. Meanwhile, initial concentration of surfactants, as a rule, was in 10–30 mg/l limits and of fats — in about 1700–2000 mg/l.

It was also clarified in the process of testing of pointed technology that in the row of cases, bad floating of fats and surfactants was observed.

Development of flotation theoretical bases as of a multistage process has led to the creation of combined flotation machines and apparatuses [38–42]. Besides, to purify wastewaters from hydrophobic easily floated compounds, for instance, fats and surfactants, combined flotation machines and apparatuses of mechanical type have been worked out. According to developed by us theoretical notions, microbubbles the most effectively interact with being floated particles. That is why to fulfill this technical settlement, a method of deep jet aeration of wastewater liquid has been developed (Fig. 10.5), the method rep-

212

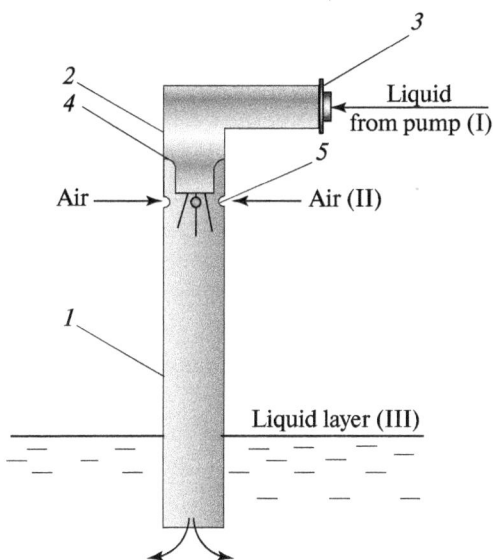

Fig. 10.5. Scheme of deep jet aeration of liquid:
1 — aerator shell; *2* — collector; *3* — sleeve; *4* — nozzle;
5 — hole for air tray

resents wastewater supply under pressure-head via jet aerator. Such device with jet system of aeration includes tube *1* which via jet supply into liquid volume is made. In the device upper part, there is chamber *2* of spread and supply of initial liquid through sleeve *3* and spread chamber of supply into tube *1* via nozzle *4*, and of sucked air — via holes *5*. Jet of liquid with gas, caught by the liquid surface, is supplied directly into volume of the liquid. Quantity of air, pumped from atmosphere, is regulated by lock valve, being fixed on the line of atmospheric air intake. In the case of heightened content of being sucked air, its reduction is reached by the way of latch armature closure. Thus, being formed mixture of waste liquid with the air is movs in restricted conditions along the tube of 1.5–2.0 m length. On the exit of tubular aerator, air dispersion goes on till tiny bubbles which form stable flotocomplexes with hydrophobic substances, for example, fats; the flotocomplexes surface-emerge into liquid upper layer. Meanwhile, coalescence of mineralized bubbles with bigger, not loaded bubbles of air proceeds. Bubbles coalesce that leads to air bubble diameter growth and, as a consequence, to growth of speed of lift of flotocomplex (particles (oil or fat drops))-bubble. Small mineralized bubbles at the expense of surface-emergence low speed don't have to reach upper border of foam layer and might be smashed down by purified liquid flow, bent in horizontal direction. For to heighten surface-emergence speed of such mineralized bubbles, it's necessary to lower the height of their lift. To embody this technical approach, fine-zoned defecating block has been developed which shelves distance between is 1–10 cm. Above-delineated technical settlements have been fulfilled in a combined flotation machine of mechanical type (Fig. 10.6) and in a flotation column with fine-zoned defecating block (Fig. 10.7).

Scheme of combined flotation mechanical machine is shown on Fig. 10.6.

A combined flotation apparatus is developed by us flotation column (Fig. 10.7) which consists of shell *1* wherein aeration chamber is fixed for supply of pulp *4* and lattice *5*.

Up chamber *2* and aligned it, floto-separating device *6* is fixed which upper part, mobile hydrophobic cap *7* is located in, and — lower part — fine-zoned separation block *6*.

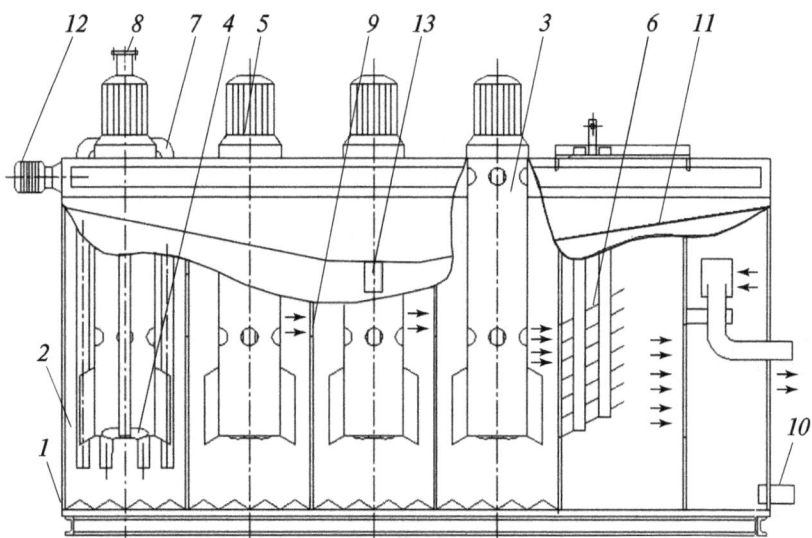

Fig. 10.6. Scheme of flotation combined mechanical machine (FCMP) with jet supply of entrance flow:
1 — shell, *2* — working space, *3* — liner, *4* — impeller, *5* — electric engine, *6* — block of fine-zoned clearing, *7* — device for jet supply of entrance flow, *8* — entrance sleeve, *9* — wall with window, *10* — exit sleeve, *11* — foam gutter, *12* — electric drive of foaming device, *13* — sleeve for foam product output

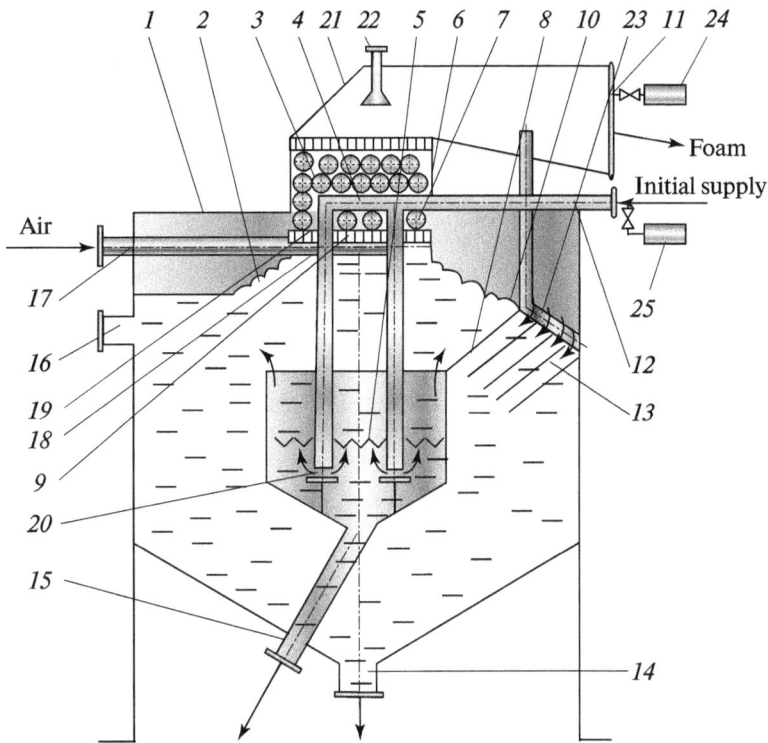

Fig. 10.7. Flotation column

In addition, hydrophobic cap is held by limiting lattices *9* which below, on inner conical part of floto-separating device *6*, corrugations *10* from hydrophobic material, fluoroplastic, for example, are made.

From external side of shell *1* of flotation column, sleeves for, respectively, foam product output *11*, pulp *12* supply, cleared liquid output (chamber product) *14*, washed liquid supply *16*, washing liquid supply *15*, air supply *17*. Fixed in lower part of flotoseparating device *6*, block of fine-zoned defecating *8* includes set of shelves *13* made in the form of vee elements (Fig. 10.8). Situated inside device *2*, jet aerators represent established vertically cylindrical tubes which upper part, there are holes *18* which upwards, nozzles *19* are fixed. In addition, under lower endings of jet aerators, reflectors *20* are situated which are fulfilled in the form of flat quadratic or round fins.

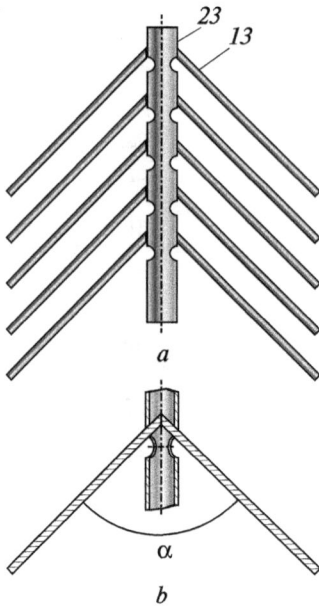

Fig. 10.8. Vee elements of block of fine-zoned decanter: *a* — block in a kit; *b* — block individual fragment

Up device *6*, there is foam tray *21* which upper part, irrigating injector *22* is fixed, accomplished in the form of filter (shower spray), and in lower part — sleeve *23* for accumulating of bubbles of gas or air in shelves *13* upper part.

Supply of initial power can be made with simultaneous supply of air under pressure from compressor *25*.

Foam product bend can be pursued also under vacuum created by pump *24*.

Flotation column works in the folowing way. Initial pulp or fine-dispersed suspension goes along entrance sleeve *12* via collector *4* into jet aerators wherein air is also pumped or supplied under pressure through holes *18*, whereat pumped air quantity is determined by speed of leaking through nozzle *19* jet of water or suspension. In addition, at the expense of underpressure, appearing at speeds of flow of water or fine-dispersed suspension higher about 15 m/s, water (or gas) suction to tiny bubbles and their contacting with suspended particles of mineral or organic nature proceed. Meanwhile, to create uniform and effective aeration across all volume of the water or fine-dispersed suspension, necessary number of aerators, as experimental investigations have shown, consistutes 4—8 per 1 m². In the case of usage of less than 4 aerators per 1 m², aeration and flotation efficiency falls down, and in the case of application of more than 8 aerators per 1 m², flotation effect doesn't grow.

Exiting from aerators with high speed, the liquid jet is additionally disperged getting on reflectors 20. Whereat, additional crushing of air (or gas) bubbles till tinier sizes, reaching ones of about 0.1—0.5 mm, and intensive sticking process of bubbles with solid phase particles and drops of hydrophobic substances, for example, of types of oils, fats and so on happen. For fuller extraction of particles of wide range size, lattice 5 with live 10—15%% section is used. This range was defined on the basis of pursued researh. At live section square of less than 15%, aeration and flotation efficiency falls down drastically, and at live section of more than 30%, the achieved possitive effect doesn't change.

Being formed flotocomplexes bubble-(particle (oil drop)) lift up, creating in flotoseparating device 6 foam layer, which contacts with hydrophobic surface of corrugations 10 of confuser inclined part and then with hydrophobic cap 7, made, for instance, from fluoroplastic balls with 5–10 mm diameter. Whereat, the confuser angle α (Fig. 10.8) in the limits of 20–70° is chosen on the basis of held investigations. At the confuser angles of less than 20°, foam product lift is much complicated that leads to drastic rise of foam time-being in the flotation column, to falling out of floated particles from it and, as a consequence, to decrease of extraction degree of target product. In the case of use of confuser with angle of more than 70°, contacting effect of inclined hydrophobic surface with foam and, correspondingly, with gas bubbles lowers and, hence, compression effect of foam layer volume lessens. The chosen limit of confuser has been checked at trials of column new sample.

At the expense of foam (foam layer) contacting with hydrophobic materials of denoted shape, intensive coalescence (sticking) of gas bubbles with each other and, as a consequence, foam layer lessening in a volume and raise of target product concentration in a foam happen. Then, foam layer, passing through lattices 9, gets into foam tray 21 where it's exposed to irrigation with water supplied via irrigating injector 22. In addition, washing out of hydrophilic and badly held in a foam particles, which get into aeration zone of device 2, takes place. Whereat obtained foam product is inadvertently dropped along inclined tray 11 or is sucked with the use of vacuum-pump 24.

Cleared liquid (chamber product) bend goes on in floto-separating device 6 wherein foam product is concentrated on account of gas bubble coalescence. Cleared liquid meanwhile passes additional cleaning by the way of defecating at slow flow between shelves 13 of block 8. At defecating in fine layer of 2–50 mm height, separation of fine bubbles as loaded with particles or oil drops as not loaded, which then accumulate in upper part of vee elements, representing shelves 13, proceeds.

Being accumulated bubbles then at the expense of lifting force, also implied by airlifting effect, are bent via sleeve 23 into foam product, situated in tray 21. Cleared liquid (chamber product) after fine-zoned defecating block is outputted from flotation column through sleeve 14.

Experimental data on efficiency of wastewater purification from fats with the use of Praestol 655 qua a reagent at the use of flotation machines of various types are presented in Table 10.7.

Data on Praestol 655 dose influence on wastewater purification efficiency from fats in flotation column apparatus with working volume of 1.5 m^3 are presented in Table 10.8.

Table 10.7. Efficiency of wastewater purification on extraction of fats in various machines with and without flocculant Praestol 655 addition (Praestol dose is 18 mg/l)

Type of flotation machines and apparatuses	Fats concentration, mg/l		
	Initial	Ending	
		Without flocculant addition	With addition of flocculant Praestol
Mechanical FCMO	1500—2000	40—60	20—30
Pneumatic PFM	1500—2000	50—80	35—60
Pressure-head	1500—2000	30—40	20—25
Jet	1500—2000	60—80	30—50

Table 10.8. Praestol 655 dose influence on wastewater purification efficiency from fats in flotation column apparatus with working volume of 1.5 m³ (concentration of fats in initial water is 1227 mg/l; aeration intensity of wastewaters is 0.5 m³/m² min)

Praestol dose 655 (mg/l)	Flotation time, min	Fats concentration in purified water, mg/l	Purification efficiency, %
0	5	156	43.9
3	7.5	73	59.4
6	10	69	71.6
9	12.5	51	81.6
12	15	37	86.7
15	17.5	22	92.1
18	20	19	93.2
21	22.5	18	93.5
24	25	18	93.5
27	30	1.8	93.5

Table 10.9. Dependence of efficiency of wastewater purification from surfactants versus flotation time in flotation column apparatus (working volume of 1.5 m³)

Flotation time, min	Surfactant concentration, mg/l		Purification efficiency, %
	In initial water	In purified water	
2. 5	10.6	0.64	94.0
5	6.5	0.25	96.2
7.5	6.5	0.27	95.8
10	8.0	0.32	96.0
12.5	8.0	0.36	95.5
15	6.5	0.24	96.3
17.5	10.6	0.43	95.9
20	10.6	0.58	94.5
22.5	10.6	0.52	95.1

Dependence of efficiency of wastewater purification from surfactants versus flotation time in flotation column apparatus is demonstrated in Table 10.9.

Experimental data, presented in Tables 10.7–10.9, point out possibility of achievement of purification high degree by the way of flotation technique usage with application of various types of flotation machines and apparatuses, including those of column type.

10.3 DEVELOPMENT OF FATTY WASTEWATER FLOTATION PURIFICATION TECHNOLOGY OF AGRO-INDUSTRIAL ENTERPRISES

Agro-industrial complex development in RF is naturally accompanied by additional burden on environment by different wastes, including those dropped by wastewaters. The analysis of wastewater content of such productions is connected, as a rule, with the presence of big quantity of fats, suspended substances, surfactants and other pollutions. Depending on the sources of formation of wastewaters, the content of denoted pollutions can vary but qualitative content is approximately preserved. Let consider wastewater content of packing industry, whereon the wastewaters form after equipment wash (Table 10.10).

Analysis of data presented in Table 10.10 points out the necessity of effective extraction from wastewaters, first of all, of fats and suspended substances because of their pronounced content. Our experience of similar wastewater purification says about the feasibility of usage of flotation and other accompanying processes, for instance, of defecating and filtration.

Efficiency check of such approach was made on lab installations. As a result of lab trials, the results were obtained which values are shown in Table 10.11.

Data presented in Table 10.11 show that obtained results meet all requirements claimed for purified water quality.

On the basis of got data, technological scheme of wastewater cleaning of above-denoted industries has been developed.

According to technology proposed by us (Fig. 10.9), wastewaters after equipment washing gather in existing accepting-collecting container (Fig. 10.9, stance *1*) and next supplied by pump into flotodecanter, denoted in technological scheme as stance *2*. Flotodecanter is equipped with chamber of mixing and flake-formation. The reagents given into flake-formation chamber are: reagent *1*, coagulant, flocculant. As a result of reagent treatment, dissolved substances are transferred into sediment with the formation

Table 10.10. Initial water content and requirements for water purification quality by stages for packing plants

Indicators	Measurement unit	Wastewaters		
		Before cleaning	After local mechanical purification	After mechanical and biological cleaning
Temperature	°C	18–25	–	10–22
Transmittance according to type	cm	0.5	1	8
Suspended substances	mg/l	2000	500	50
Fats	mg/l	1000	50	0
Smell	point	5	–	2
Color	–	red-brown	–	colorless
pH	–	6.5–8.5	–	7–8.5
Hardness, general	mg-equiv/l	10	–	10
Hardness, carbonaceous	mg-equiv/l	10	–	–
Salt content	mg/l	1500	1500	1000
Ca^{2+}	mg/l	75	–	–
Mg^{2+}	mg/l	50	–	–
Cl^-	mg/l	900	–	500
SO_4^-	mg/l	500	–	–
Fe_{common}	mg/l	20	–	1
COD	mgO/l	2000	1000	50
BOD_5	mgO_2/l	800	400	30
Nitrogen general	mg/l	150	–	–
Phosphorus (calculated per P_2O_5)	mg/l	60	–	–
NH_4^+	mg/l	30	–	5
NO_2^-	mg/l	0.02	–	–
NO_3^-	mg/l	0.05	–	–
Chlorine, active	mg/l	0	1.5	1.5

Table 10.11. Results of laboratory trials

Nomenclature of indicators, measurement units	Value of indicators		
	Initial water	Water after purification	Defect action level*
Ammonium-ion, mg/dm^3	47.5	0.5	0.5
Anionic surfactants, mg/dm^3	< 0.025	< 0.025	0.5
Suspended substances, mg/dm^3	426.0	10.0	10.75
Hydrogen indicator (pH), points	8.3	8.0	6.5–8.5
Iron, mg/dm^3	10.89	< 0.04	0.1
Fats, mg/dm^3	38.6	0.9	—
Mineral oils, mg/dm^3	1.732	0.049	0.05
Nitrite-anions, mg/dm^3	0.20	< 0.02	0.08
Sulphate-anion, mg/dm^3	59.0	34.2	100.0
Phosphates, mg/dm^3	0.55	0.12	0.2
Dry residue, mg/dm^3	286.3	156.3	1 000.0

*Federal Fishing Order N.20, January 18, 2010 "On Adoption of Norms of
Water Quality of Ponds of Fishing Value, Including Norms of Action Level of
Harmful Substances in Waters of Ponds of Fishing Value"

of big flakes. The flakes fall out into sediment in the defecating zone. For
to intensify the process, the defecating zone is equipped with fine-zoned
module. In the flotation part, the major part of left oils (about 90%), fats,
surfactants as well as other pollutions, treated with reagents, are removed
from the wastewater. After flotation cleaning, the water is gathered in intermediate container (stance 4) wherefrom the water is supplied for polishing
to pressure-head mechanical (stances 5.1, 5.2) and sorption filters (stances
6.1, 6.2). Qua a fill in the filters, a zeolite with weakly pronounced sorption
properties and activated carbon of AG-3 brand, which securely provide for
polishing high degree from fat- and oil-containing wastewaters, are used. To
intensify mechanical filtering process, coagulant input before mechanical
filter (5.1) is provided. Such approach allows to pursue the process of contact
coagulation on the surface of grains of the filtering fill. At contact coagulation, coagulant starts to contact with earlier formed flakes of the coagulant,
suspended substances and filter grainy fill that much raises the degree of extraction of fine-dispersed and colloidal pollutions. In the process of work of
the mechanical filters, pollutions are accumulated across filtering fill volume. Regeneration of mechanical filters is pursued by reverse wash by the
purified water from container 7. Washed waters after regeneration of filters
(5.1 and 5.2) are dropped into accepting container (stance 1).

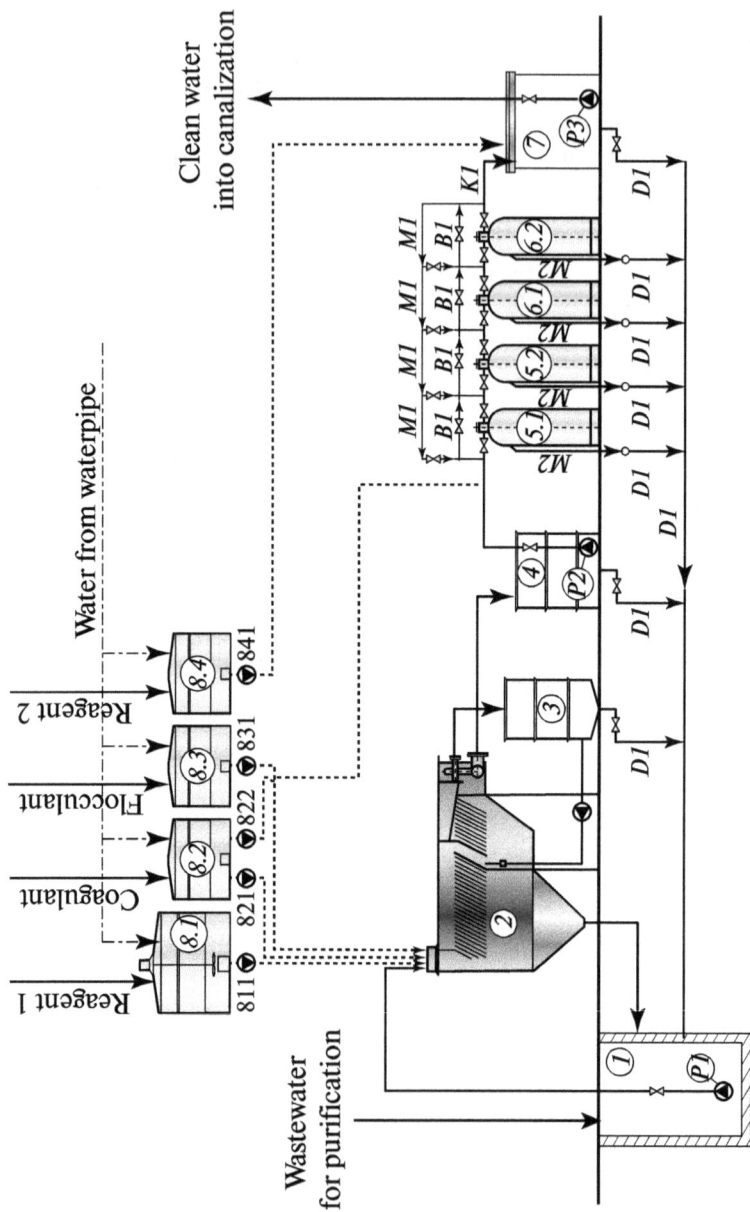

Fig. 10.9. Principled technological scheme of purification of fat-containing wastewaters

a   *b*

Fig. 10.10. Scheme of aggregates of pollution particles in the form
of aero-floccules and flakes from pollution particles:
a — air-floccules; *b* — flakes from dirt particles

After polishing stage, the water is supplied into clean water container (*7*). Correction of pH is made in the container. The water with pH 6.5–8.5 from the container is dropped with the help of loading pump (stance P*3*) into canalization. Purified water, being accumulated in the clean water container, is used for wash of the mechanical filters. Worked-out water after filter wash is dropped into accepting container (*1*),

Caught in flotodecanter, mud is periodically dropped into container for mineral oils (stance *3*).

Suspended substances, settled in the flotodecanter, are periodically dropped into accumulating container and then are pumped by silt-sucking machine.

Wastewater purification on this installation provides for the fulfillment till indicators, meeting norms, claimed for the dropping into a canalization.

Purification scheme, proposed for a realization, is distinguished by the stability to alternations of water flow as by quantity as well as by quality.

Purification principle is in the following. The wastewaters mix with reagents with the help of mixer with the creation of flakes of enough big size whereat in the majority of cases representing aero-floccules (Fig. 10.10*a*). These aero-floccules are extracted primarily, and after, flakes from pollution particles are being done (Fig. 10.10*b*).

The purified liquid quality meets the requirements claimed to wastewaters being dropped into ponds of fishing value.

11 PERSPECTIVES OF THERMO–FLOTATION USE IN WASTEWATER PURIFICATION PROCESSES

The usage of flotation traditional methods doesn't always give possibility to get adequate results of wastewater purification, especially, those of complicated composition. In this regard, many developers including us pursue the search of new flotation approaches allowing not only to create a high-effective aeration system but also to promote the destruction of such stable systems as wastewaters do constitute. Preliminary research made by us on thermo-flotation use for wastewater purification has shown that this method in individual cases is enough effective in comparison with other flotation approaches, for instance, pneumatic and mechanical flotations. Works [82–83] note on perspectives of thermo-flotation usage in biotechnological production, in particular, for yeast suspension thickening.

Thermo-flotation usage for wastewater cleaning from protein substances and microorganism cells also represents an interest. It's known that the heating of being purified liquid proceeds in the case of thermo-flotation that leads to the denaturation of protein substances and formation of rather big aggregates. The latter contributes to their flotation extraction.

The check of being proposed mechanism of flotation extraction of protein substances from wastewaters of biotechnological productions was held by us on installation with working volume of 1 l. Meanwhile, we used spiral with electric power of 0.7 kW as an aerator. Quality analysis of purified water was being held with the use of photo-calorimeter (by water muddiness).

The scheme of laboratory installation is presented on Fig. 11.1.

Experimental research results are shown on Fig. 11.2–11.3.

Data demonstrated on Fig. 11.2 testify that kinet-

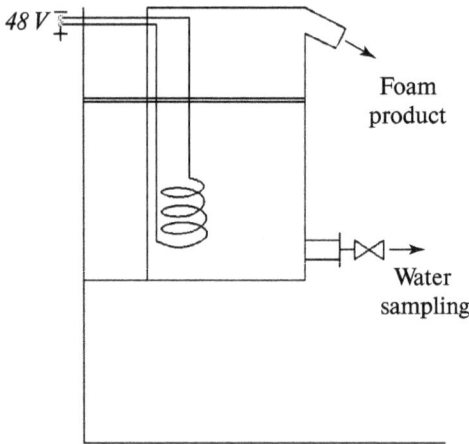

Fig. 11.1. Laboratory installation scheme

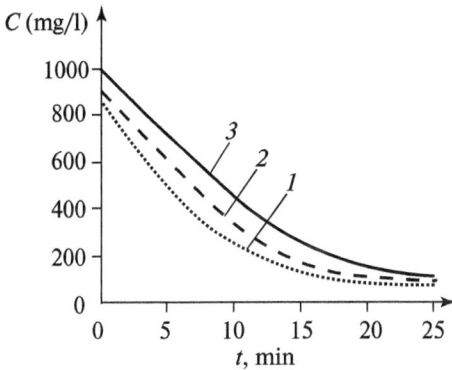

C (mg/l) axis with values 1000, 800, 600, 400, 200, 0
t, min axis with values 0, 5, 10, 15, 20, 25

Fig. 11.2. Thermo-flotation time impact on kinetics of wastewater clearing with initial concentration of suspended substances as
1 — 850 mg/l; 2 — 900 mg/l;
3 — 1000 mg/l

ics of wastewater clearing is influenced by suspended substance content in a water. In the processes of thermo-flotation purification of wastewaters, containing various biopolymers, for instance, proteins, nucleic acids and so on as well as microorganism cells and their fragments, the aggregation of pointed substances goes on at the expense of thermal impact. Meanwhile, being formed floccules together with suspended compounds contain tiny bubbles. It's more correct to call such floccules as aero-floccules. Aero-floccule formation process proceeds rather quickly, as a rule, for 10–15 minutes. We can think in a whole that thermo-flotation process goes on for about 20 minutes in average.

Special investigations, held on definition of organic compound content in liquid phase after thermo-flotation pursuing, point out increase of the compound content in the liquid phase. Probably, it's connected with the extraction of intracell biopolymers from microorganism cells.

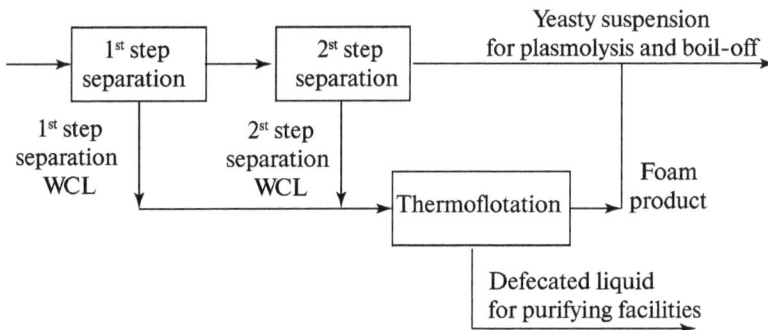

Fig. 11.3. Scheme of separational extraction of biomass of yeast
Candida kind with cleaning of wastewaters (WCL)
by thermo-flotation

Defecated liquid
after thermoflotation

Cooled defecated liquid suspension ← [diagram box] → Warmes yeasty

Yeasty suspension,
given for 1st step separation

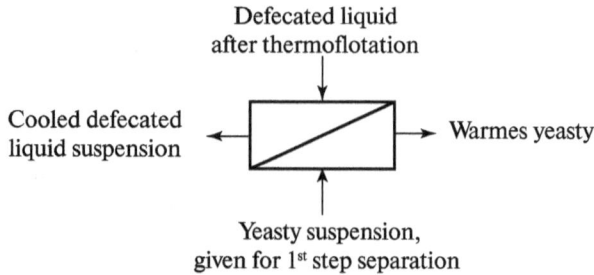

Fig. 11.4. Scheme of preliminary heating of yeast suspension

To extract biopolymers from the liquid, it's advisably to apply coagulants and flocculants. Pursued search in this direction has shown that it's sufficient in the majority of cases to apply flocculants only, as a rule, of cation type, for instance, like Praestol. Whereat, flocullant doses usually do not exceed 25–40 mg/l. Investigations of thermo-flotation cleaning of wastewaters, pursued by us, confirm this interval of doses of flocculant of Praestol type.

Value of energetic expenditures has big significance at thermo-flotation accomplishment. The results of thermo-flotation purification trials have shown that feasibility of thermo-flotation usage can be substantiated in the condition of application of thermo-processing in the course of main technology, for instance, during separational extraction of yeast biomass with preliminary thermo-treatment at 80–90 °C. Such technology of yeast biomass extraction leads to obtaining of hot flows ($T = 70$–80 °C) of worked-out culture liquid (WCL), being formed at separation of yeast suspension warmed till 80–90 °C.

Yeast biomass extraction with preliminary thermo-treatment till 80–90 °C is made in the majority of cases by typical scheme with the use of double-step separation. The possibility to use thermo-flotation was checked by the way of WCL after-purification (polishing) by thermo-flotation (Fig. 11.3).

It's worth to note that presented on Fig. 11.3 scheme of separational thickening can be exploited as with preliminary heating of yeast suspension as well as without the heating. In the first case, technical-economical indicators of the thickening process are higher than without preliminary heating of yeast suspension. In this regard, technological scheme of separational thickening of yeasts with the preliminary heating more often was used in the plants of nutritional yeast production.

Negative impact of heightened temperatures of cleared liquid on biological purification quality was further settled by the way of thermo-exchanger usage (Fig. 11.4) wherein cleared liquid warmth was used for yeast suspension heating. This approach was allowing to lower general energetic expenditures on the yeast suspension heating.

The scheme of the preliminary heating of yeast suspension is demonstrated on Fig. 11.4.

Such approach of preliminary heating of yeast suspension allows to economize till 30–40% of common thermo-expenditures on yeast suspension heating before its supply for separation of 1st step.

Thermo-flotation usage in other cases, namely, without usage of thermo-treatment in the major technology, should be preliminary substantiated. In energy-saving conditions which are followed by the majority of industrial productions the question of thermo-flotation usage can be settled at the accomplishment of well-developed technical-economical substantiation.

12 DEVELOPMENT OF WATER–USE RECURSIVE SYSTEMS WITH THE USAGE OF FLOTATION TECHNIQUES

The earliest works on the creating of factually functioning water-use recursive systems with our participation were held more than 30 years ago on enterprises of mining-chemical and microbiological industries, in particular, on nutritional yeast plants. In the process of the technology of yeast cultivation, basing on mineral oil paraffines, with the use of recursive water-use, it's been clarified that to embody the pointed technological process, the wastewater purification degree, being reached after complete biological cleaning in aerotanks, is sufficient enough. Whereat, biologically purified water contains 50–200 mg/l of a phosphorus, 80–6000 mg/l of a nitrogen, 5–20 mg/l of an iron, 50–200 mg/l of active silt biomass, 1–3 mg/l of an oxygen. Biologically purified water before its return onto cultivation stage has pH about 5.0 and temperature of 15–250° C. As a result, at full return of biologically purified water to yeast cultivation stage, the biomass output from used substrate constitutes about 100%, whereat raw protein content in a ready product is 63–65% and of an ash– about 7–8%. To compare, let denote that without biologically purified water return onto the yeast cultivation stage, the biomass output from used substrate constitutes 91–93% at about same content of raw protein and ash in the ready product. According to our data, the biomass output rises at the use of biologically purified water on the yeast cultivation stage is observed at the expense of presence of additional nutritional sources.

This example shows that even to grow microorganisms, drinking-purpose water doesn't have to be used but, for example, biologically purified wastewater can be used. Industrial introduction of this measure has allowed to cut drastically the rate of water of drinking and of technical purposes that has given noticeable effect of and has improved ecological situation which is connected with dropping of biologically purified waters into open ponds.

On the enterprises of mining-chemical raw, in particular, in the process of enrichment of phosphate raw, various approaches for water purification have been applied beginning from defecating in tailing dumps till the usage of biological purification in the aerotanks.

The intensification of the most frequently used defecating process in the majority of cases was pursued with the use of coagulants and flocculants. Last time, we have used for these purposes aluminum-containing coagulants and Praestol and protein-containing reagents on active silt basis qua a flocculant.

Biological purification usage on enterprises of phosphate raw concentrating is complicated because of low content of organic dirt in a wastewater and hence, hasn't got a wide spread.

The introduction of recursive systems on mining-chemical raw enterprises haven't lowered technological indicators and has given essential effect on fresh water intake saving.

Further works in this direction are held by us in a whole for further expansion of the recursive system usage on other industrial branch enterprises and are connected in the majority of cases with the increase of the depth of wastewater cleaning and the choice of the most rational way of their conditioning and disinfection.

The other example of the usage of the recursive systems of water use, which is faced rather frequently in practice, is connected with a carwash. According to RF norm documents, carwash must be held in the regime of closed water use with washing water purification on local purifying facilities (LPF).

LPF of our know-how include 3 sectional sand trap, dirty water container with plunging pump, dirt-oil-catcher, floto-machine of pneumatic or pressure-head type, polishing filter.

LPF work principle is in the following.

Wastewaters after carwash, delivered into 3 sectional sand trap, are purified from sand particles of size of no more than 0.15–0.20 mm. Then, drainages, purified from sand big fractions, are delivered of its own accord into dirty water container wherefrom they are directed after preliminary defecating with the help of plunging pump into dirt-oil-catcher with a fine-zoned clearing block.

In the process of wastewater passing through special blocks of fine-zoned defecating, dirt-oil-catchers, fixed in the 1st chamber, the water separation from fine-zoned suspended particles of less than 0.15 mm size goes on as well as from drops of mineral oil products. Meanwhile, the suspended particles fall out into sediment which is removed by valves opening for sediment sink, and oil products emerge up forming oil film which is deleted periodically by means of a level regulator via special gutter with film bend into the capacity for foam product collection.

After a dirt-oil-catcher, partially purified wastewaters are inputted of its own accord into pneumatic type floto-machine where water purification from fine-dispersed drops of oil products proceeds at the expense of their sticking with air bubbles with further lever into upper foam layer which is removed periodically with the help of level regulator and is directed into capacity for caught oil product collecting.

Cleared liquid by gravity flow is supplied from flotomachine into an overshot filter where in the process of filtration, via grainy fill layer, water polishing from residual oil product, existing in fine-dispersed or dissolved state, goes on.

Water, purified from sand and oil products, after the overshot filter enters by gravity filter the clean water container wherefrom with the help of the plunging pump it is directed for next repeating use on a wash.

Purified water quality, given on the wash of automobiles, usually must meet the following conditions.

Oil product concentration mustn't exceed:
- 15 mg/l for motor cars;
- 20 mg/l for trucks.

Suspended substances:
- 40 mg/l for motor cars;
- 70 mg/l for trucks.

Typical scheme of wastewater purification developed by us is demonstrated on Fig. 12.1.

Similar schemes of wastewater purification were used by us mostly before year 2000. Various variants of the above-considered typical scheme were used by us further, taking into account the specificities of industrial introduction of the object.

Whereat, the modification touched either the increase of depth of wastewater purification or the use of devices for water disinfection. Purification efficiency increase was reached by the use of clearing additional blocks being installed inside flotation machine shell and in some cases by the use of other type flotomachines, for instance, mechanical and flotation columns with aeration jet system. Flotation machine FKMO-0.5 of mechanical type developed by us has been applied in the series of objects of Moscow region. In the case of wash aggregates usage of Karcher brand, additional filters have been used (Fig. 12.2).

Together, filters with carbon fill and water disinfection devices also can be applied in the technological scheme.

Special attention is deserved for the possibility of usage of flotation columns (Fig. 12.3 A, B) where rather high probability of particle collision with gas bubbles is provided leading to particle sticking and later on to preservation of mineralized air bubbles. Relative speeds of bubbles and particles of dirt in columns alternate in 10–12 cm/s limits that according to the data of Czech researcher Dedek, creates optimal conditions for their sticking. The blend-

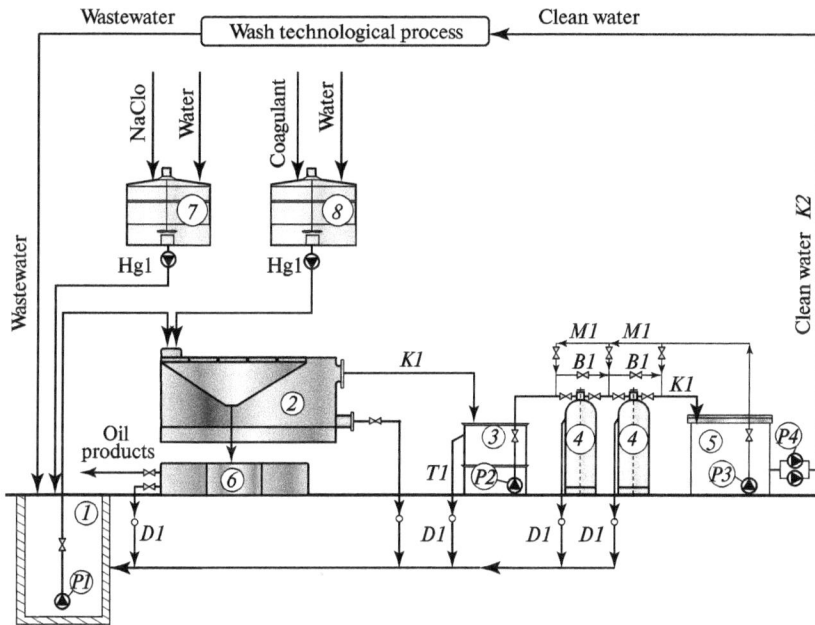

Fig. 12.1. Principled scheme of purification facilities of carwash:
1 — pit; *2* — flotomachine; *3* — intermediate reservoir;
4 — filters; *5* — clean water container; *6* — collector of oil products;
7 — tank for natrium hypochlorite preparation;
8 — tank for coagulant preparation; *P1, P2, P3, P4* — pumps;
Hg1 — pumps-dispenser; *K1* — being purified water stream;
K2 — purified water stream; *M1* — purifying water stream;
B1 — bypass pipeline; *T1* — transfusion stream; *D1* — drain waters

ing devices, generating inertial forces, causing particle detachment from the bubbles, are absent in the columns; this rises the preservation of the flotocom-plexes (pollution particle)-bubble.

There should be also related to advantages of counter-current columns: low power-consuming, small space necessary for installation, constituting about 20–30% of square, occupied by standard machines of the same productivity, as well as wide possibility to use the processes of secondary mineralization in the foam layer to rise the efficiency of sticking of bubbles with dirt particles.

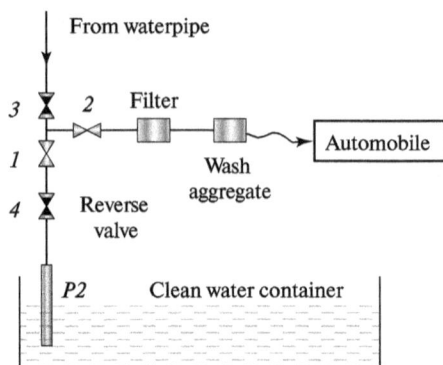

Fig. 12.2. Scheme of water supply onto wash aggregate

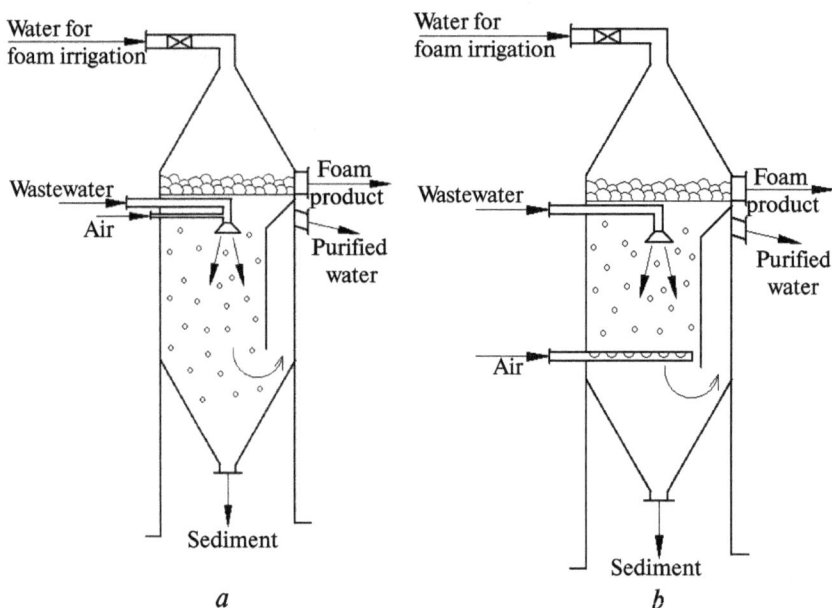

a   *b*

Fig. 12.3. Scheme of flotation column for purification of wastewaters
and fine-dispersed pulps from hydrophobic pollutions:
a — straight-through regime of aeration;
b — counter-current regime of aeration

To confirm the pointed advantages, trials with the use of experimental and experimental-industrial samples with counter-current regime of work were held (Fig. 12.3 *b*).

**Table 12.1. Flotation time influence on oil product
residual concentration in purified water
(oil product concentration in initial water is 27.8 mg/l;
wastewater aeration intensity is 1.1 m³/m² min)***

Number N.	Flotation time, min	Mineral oil product concentration in purified water, mg/l	Purification efficiency (%)
1	5	15.6	43.9
2	7.5	11.3	59.4
3	10	7.9	71.6
4	12.5	5.1	81.6
5	15	3.7	86.7
6	17.5	2.2	92.1
7	20	1.9	93.2
8	22.5	1.8	93.5
9	25	1.8	93.5
10	30	1.8	93.5

**Table 12.2. Averaged values of oil product concentration
in purified water depending on time of wastewater flotation
in flotation column with working volume of 1 m³***

Number N.	Flotation time, min	Mineral oil product concentration in initial water, mg/l	Residual concentration of oil products in purified water, mg/l	Purification efficiency (%)
1	5.7	32.4	16.6	48.8
2	10.1	30.1	13.1	56.5
3	14.5	28.9	6.9	76.1
4	20.2	31.6	4.8	84.8
5	23.9	29.8	3.3	88.9
6	30.6	31.2	3.4	89.1

*Trials were held in periodical regime.

Testing the the efficiency of flotation column apparatus usage was being held at the purification of mineral oil containing wastewaters. In addition, on the stage of test trials, flotation column sample with working volume of 15 l was used. Test trial results are shown in Table 12.1.

Analysis of data presented in Table 12.1 testifies the achievement of rather low oil product concentration in the purified water, not exceeding 1.8 mg/l, at flotation time of oil-containing drainages of no more than 22.5 min. Such indicators in wastewater purification practice are considered to be high, they point out the perspectives of flotation column apparatus usage for wastewater purification from hydrophobic pollutions, for instance, mineral oil products.

To check flotation column technique efficiency in experimental-industrial conditions, the trials on apparatus of 1 m3 working volume were being pursued. Column apparatus of the same design as trial sample was used. Results of these trials are demonstrated in Table 12.2.

Data, presented in Table 12.2, tell on the confirmation of testing results obtained earlier in a laboratory environment. Some difference in values of oil product residual concentration in the purified water, outputting from lab and experimental-industrial samples of the flotation column, is implied by large scale difference of delineated apparatuses.

We can state, evaluating the data, obtained in a whole, that column flotation apparatuses don't lose in technological efficiency to flotation machines of typical designs [5] that offers possibility to apply them in the practice of wastewater cleaning from hydrophobic pollutions, including also from mineral oil products. The application of such apparatuses is especially effective in the case of the necessity of extremely compact disposition of equipment because of industrial limited spaces.

As research results have shown, the special interest for practical usage is in the flotation column with jet and injection aerators. In such flotation columns, those conditions are realized at most that are mentioned above, including heightened content of fine-dispersed gas and creation of favorable conditions for gas bubble contacting with dirt particles.

Acquired data are easy interpreted also from the point of multistage flotation theory that has been being developed by us last time [50]. Gas bubble size decrease leads to the growth of constant, characterizing the probability of the sticking of dirt particles with a bubble. Constant value grows also at the increase of the probability of the capture of pollution particle by gas bubble that is exactly observed in column flotation apparatuses which work in counter-current regime, i.e. at meeting-lane movement of liquid (aqua) and gas phase. It's worth to note that the growth of probability of dirt particle capture in the column flotation apparatuses increases, as the results of our special investigations show, by about 1.5 times in comparison with this parameter values in typical flotation machines. Namely emphasized circumstances, first of all, appear in the majority of cases as defining ones at the choice of flotation apparatus type for wastewater purification from hydrophobic pollutions, including also those at the usage of water recursive systems. For example, in the case of the development and introduction of compact systems of recursive water use, there is no doubt in the feasibility of flotation column usage.

13 FLOTATION THICKENING OF EXCESSIVE ACTIVE SILT

13.1 INTENSIFICATION OF ACTIVE SILT THICKENING

Intensification of flotation thickening process of active silt suspension depends also on its preliminary processing, for example, with the use of chemical reagents, with various physical impacts, in particular, heating. At physical-chemical processing of active silt suspension, processes of aggregation of microorganisms go on, in particular, under the effect of polysaccharides, being extracted in such processing. Table 13.1 shows values of polysaccharide concentration in liquid phase of heated till 85 °C suspension of active silt. It's observed meanwhile that polysaccharide concentration grows by 3–5 times at pH raise till 8.5 and, together, aggregation effect is seen.

Table 13.1. Influence of thermo–reagent treatment of active silt suspension on polysaccharide concentration in the suspension liquid phase (active silt concentration in the suspension is 0.5 %)

N.	Processing parameters			Polysaccharide concentration, mg/l	Clearing efficiency (%) (Turbidity measurement by Photoelectric colorimeter)
	t, °C	pH	Holding time, min		
1*	20	7.2 (initial)	–	11	54.3
2	40	7.2	5	35	62.8
3	80	7.2	5	52	73.1
4	85	7.2	5	54	75.8
5	90	7.2	5	53	74.6
6	20	8.0	5	9	52.3
7	20	8.5	5	9	51.7
8	85	8.5	5	8	50.9
9	85	8.5	10	35	61.6
10	85	8.5	30	31	60.7
11	85	8.5	45	31	61.1
12	85	8.5	60	9	50.8

* The control.

Table 13.2. *Averaged data on influence of dose of cation–active floc-culants on efficiency of active silt flotation thickening*

Flocculant kind	Flocculant dose, kg/tn	Active silt concentration in foam, % mass	Active silt concentration in liquid phase, % mass
Praestol 644	1.5	3.1	0.13
	2.5	3.5	0.11
	5.0	4.3	0.10
	7.5	4.5	0.8
	10.0	4.6	0.8
Praestol 852	1.5	3.4	0.11
	2.5	4.2	0.09
	5.0	4.8	0.07
	7.5	4.9	0.03
	10.0	4.9	0.03

The obtained data point out the expediency of such physical-chemical treatment as one of the methods of intensification of active silt flotation thickening. Another, in the majority of cases more effective way is the preliminary treatment of active silt suspension with reagents, particularly, with cation-active flocculants of Praestol 644 and 852 types. The data of denoted flocculant influence on the process of flotation thickening of active silt suspension are shown in Table 13.2.

The analysis of data presented in Table 13.2 testifies that cation-flocculant Praestol 852 usage leads to the best results than flocculant Praestol 644 usage at the same doses. In this regard, flocculant Praestol 852 has been recommended to further usage in the practice of flotation thickening of active silt suspension.

Thus, pursued investigations have demonstrated that flotation apparatuses with aeration pneumatic-hydraulic system can be used in the practice of active silt flotation thickening, and it's feasible to use cation flocculant Praestol 852 for this process intensification.

13.2 THICKENING OF ACTIVE SILT SUSPENSION WITH THE USE OF FLOCCULANTS

Usage of flocculants for the thickening of fine-dispersed suspensions, including active silt, wastewaters, have got rather wide spread in the world practice [6]. However, the mechanism of interaction of flocculants and dirt particles and, hence, their usage possibilities haven't been yet fully disclosed so far.

In this regard, research on flocculant usage represents interest not only for deletion of suspended particles but of dissolved ones from liquid phase as well. It's worth to note that flocculant action efficiency extensively depends on preliminary addition of coagulants into being cleared suspension. Incidentally, research of flocculant impact on suspension thickening process was held with preliminary addition of coagulants as well as without their usage. Mainly vitriolic iron and sulphuric acid were used qua coagulants, and Fennopol (production of Finland) and VPK-402 (Russia) qua flocculants. Samples of active silt, taken from purification facilities of Pulp and Paper Factory, town Summa (Finland), Kirish Biochemical Plant (Russia) and model systems, containing organic compounds were subjects for the research.

Research of reagent impact on process for clearing of suspension of active silt, taken on Summa purification facilities, was held with the use of pressure-head flotation, but with the use of defecating in cylinders — on purification facilities of Kirish Biochemical Plant.

Testing of vitriolic iron ($FeSO_4$) and Fennopol supplements on samples of active silt, taken on purification facilities of Summa town, was pursued during mixing, and then worked out with reagents, active silt suspension was mixed with working liquid (dispersion water) in ratio 1:1. Vitriolic iron ($FeSO_4$) in the form of liquid was used qua coagulant, the vitriolic iron was obtained from Pulp and Paper Factory, Summa town, Fennopol — also from Pulp and Paper Factory, Summa town, — qua flocculant.

Assessment of impact of reagents was pursued on the height of foam layer, being formed during flotation, by COD (Chemical Oxygen Demand) and liquid phase turbidity.

Table 13.3 shows conditions of active silt suspension treatment with reagents and achieved thus separation effect of active silt suspension.

Data presented in Table 13.3 indicate the efficiency of being used reagents.

It's been established in the case of investigations pursued on samples of suspension of active silt collected on Kirsh Biochemical Plant facilities that one of the effective reagents is cation flocculant VPL-402 added jointly with vitriolic iron or with sulphuric acid. Research results are demonstrated in Tables 13.4 and 13.5.

The obtained data on foam layer compression, being formed at active silt flotation with addition of vitriolic iron and Fennopol, show (Fig. 13.1, 13.2) that foam layer volume (height) decreases by 1.5–2.0 times in comparison with the control (without reagent addition) at vitriolic iron concentration of 0.25–0.35 mg/l and Fennopol concentration of 1–10 mg/l.

Table 13.3. Reagent treatment influence on active silt suspension separation by flotation (flotation time 10 min, active silt to dispersing water ratio 1:1)

N	Active silt suspension quality	Being added reagent quantity		pH of active silt suspension	Separation efficiency		
		FeSO$_4$ In the form of solution, ml	Fennopol (Summa) 0.01% solution, ml		Foam layer height, cm	Turbidity	COD (mg/l) of liquid phase
1	Active silt after stabilization during 1 hour	–	–	6.93	5.0	48	170
2	Active silt after 6 hours of being in anaerobic conditions	–	–	6.94	5.2	45	260
3	Active silt after hourly aerobic stabilization	0.08	–	5.20	3.0	43	80
4		0.08	1.0	5.10	3.2	42	65
5		0.08	2.0	4.68	3.0	22	–
6		–	–	6.95	–	–	–
7		1.0	–	5.1	4.5	30	–
8		0.25	–	6.0	4.0	28	–
9		0.10	–	6.5	4.0	25	–

Table 13.4. Vitriolic iron impact on thickening process of active silt suspension by defecating at pH various values (ADS (Absolutely Dry Substances) concentration in initial suspension is 0.92 %)

N	Regimes of treatment of active silt suspension		Preliminary acidification of active silt suspension by sulphuric acid (pH units)	Microbe biomass concentration (% ADS)	
	Iron sulphate rate, %	Mixing		In sediment	In cleared liquid
1	–	–	6.9	1.44	0.41
2	0.01	mixer, 5 min	6.9	1.62	0.22
3	0.05	mixer, 5 min	6.9	2.03	0.18
4	0.1	mixer, 5 min	6.9	2.32	0.12
5	0.01	pneumatic, during 5 min	6.9	1.56	0.37
6	0.05	pneumatic, during 5 min	6.9	1.98	0.20

N	Regimes of treatment of active silt suspension		Preliminary acidification of active silt suspension by sulphuric acid (pH units)	Microbe biomass concentration (% ADS)	
	Iron sulphate rate, %	Mixing		In sediment	In cleared liquid
7	0.10	pneumatic, during 5 min	6.9	2.36	0.12
8	0.01	pneumatic, during 5 min	5.0	1.58	0.36
9	0.05	pneumatic, during 5 min	5.0	2.10	0.22
10	0.10	pneumatic, during 5 min	5.0	2.38	0.11
11	0.01	pneumatic, during 5 min	4.0	1.64	0.33
12	0.05	pneumatic, during 5 min	4.0	2.16	0.20
13	0.10	pneumatic, during 5 min	4.0	2.40	0.11
14	0.01	pneumatic, during 5 min	3.5	1.60	0.32
15	0.05	pneumatic, during 5 min	3.5	2.15	0.18
16	0.10	pneumatic, during 5 min	3.5	2.44	0.10

Table 13.5. Cation flocculant VPK–402 impact on active silt suspension separation by defecating during 1 hour (ADS (Absolutely Dry Substances) concentration in initial active silt suspension 0.84 %)

N.	Reagent concentration			Microbe biomass concentration (% ADS)	
	Iron sulphate, %	Sulphuric acid, mg/l	VPL-402, mg/l	In sediment	In cleared liquid
1	0,05	–	4	2.21	0.18
2	0.05	–	8	2.24	0.13
3	0.05	–	12	2.27	0.12
4	0.05	–	16	2.19	0.07
5	0.05	–	20	2.18	0.07
6	–	0.6	4	2.63	0.10
7	–	0.6	8	2.54	0.10
8	–	0.6	12	2.48	0.09
9	–	0.6	16	2.50	0.06
10	–	0.6	20	2.52	0.06

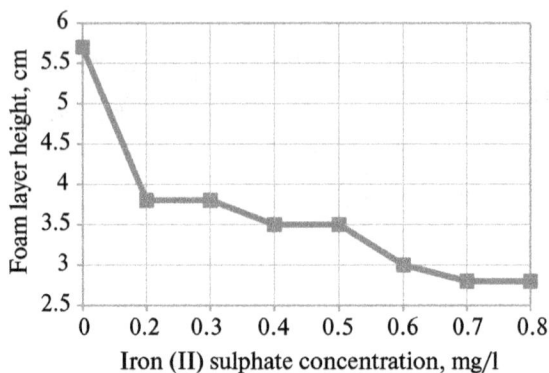

Fig. 13.1. Dependence for height of foam layer of floated active silt versus being added iron sulphate rate

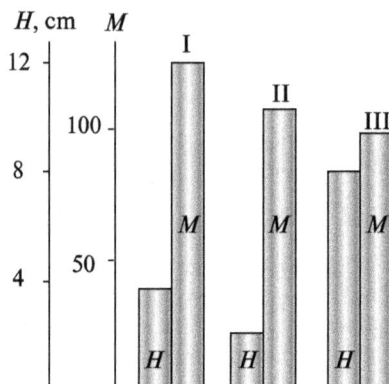

Fig. 13.2. Impact of Fennopol in combination with iron sulphate on efficiency of flotation division of active silt

Meanwhile, also turbidity value of being defecated liquid phase decreases by 3–5 times that points at content decrease of suspended and dissolved compounds in liquid phase.

At vitriolic iron ($FeSO_4$) concentrations of more than 0.3–0.4 mg/l, more pronounced effect of foam layer compression is reached, but at it, turbidity value raises that is unwilling also in economic terms and probably inadvisable because thus vitriolic iron rate grows.

There is a practical and scientific interest to compare efficiency of active silt suspension treatment with vitriolic iron in combination with Fennopol and the treatment with sulphuric acid (Fig. 13.3, 13.4).

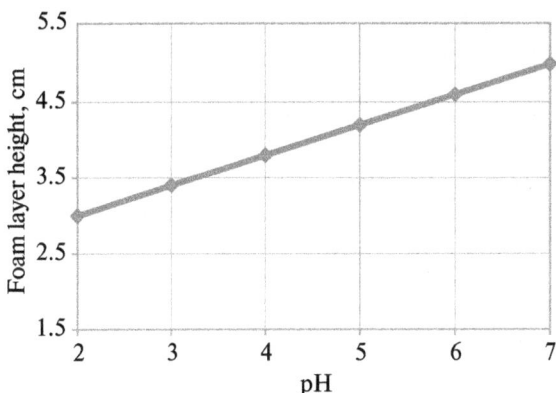

Fig. 13.3. Dependence of height of foam layer of floated active silt versus pH value

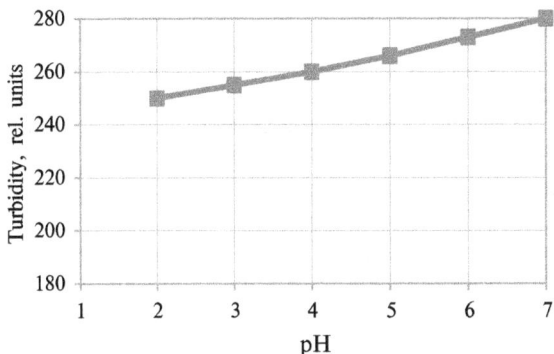

Fig. 13.4. Dependence of cleared liquid turbidity after active silt separation by flotation versus pH value of liquid phase

Active silt treatment with sulphuric acid leads to coagulation of active silt particles and improvement of division of suspension into foam layer and liquid. Meanwhile, the foam appears to be 2.0–2.5 denser at pH 2.5–4.5 than at pH 6.6–7.0 (without the acid addition). At pH values of 2.5–4.5, cleared liquid turbidity lowers by 4–5 times in comparison with turbidity of the liquid, being obtained at active silt suspension division by flotation without sulphuric acid addition.

The obtained results on Fennopol impact on flotation process of active silt correlate with the data, obtained at investigation of VPK-402, on defecating process of active silt. In both cases, efficiency of these cation flocculants grows with the use of additional coagulants — iron sulphate and sulphuric acid. It's known that macromolecules of both flocculants carry positively charged groups, particularly, VPK402 macromolecules include groups $N+(CH_3)_2Cl$, allowing to achieve charge of rather high value of about 40 mV. Charge presence of such value points at the possibility of interaction of cation flocculants with dissolved in water organic substances, for instance, anion-active collectors of the type of carbonic acid soaps. The choice of anion-active collectors as a research object is implied by that they together with cation-active flocculants can be potentially used in the processes of active silt flotation.

For to use concrete reagent regime it's necessary to pursue investigation directly on industrial introduction object.

13.3 INTRODUCTION INTO INDUSTRY OF FLOTATION DEVICE FOR EXCESSIVE ACTIVE SILT THICKENING

Results of technological trials of pressure-head flotation skimmer for active silt thickening are shown. It's been established that flotation thickening of excessive active silt allows to prognose the obtaining of rather high degree of silt suspension thickening (by 10 and more times) and to forecast the specific productivity increase of similar type equipment by 1.5–2.5 times in comparison with the existing one.

Large quantity of excessive active silt is formed in the course of exploitation of purification facilities. Up to 3.5–4.5 million tn. of excessive active silt biomass is accumulated annually on purification facilities of wastewater biological cleaning in RF. Complex problems, connected with the specificities of active silt suspension and the difficulties of dehydration of the suspension, occur at active silt thickening.

Impact of significant number of factors drastically complicates pursuing of biochemical purification in optimal regime. Last time, some progress has been outlined in the biochemical process management, probably, because of extended development of biotechnology and, first of all, microorganism industrial cultivation. Nevertheless, difficulties of holding of biotechnological processes, including cultivation on wastewaters of microorganisms of active silt, aren't eliminated in the series of cases. It corresponds especially to the management of such complex processes as flocculation of active silt microorganisms, active silt separation from water and further thickening.

Flotation has definite advantages to other methods of thickening, its usage for excessive silt packing gets more and more spread. In addition, the flotation is used as a rule on the first stage of active silt thickening before its further centrifuging and filtering.

Ksenofontov B.S. has pursued in LLC "GosNIIsynthesisbelok" the experimental research on flotation thickening of active silts with various silt index (in 40–60 cm^3/g frames), taken from local purification facilities of biochemical production. The research has shown that microbe biomass thickening degree in being obtained foam product constitutes 4–6. To achieve higher thickening degree, it's necessary to use complementary devices for packing of foam layer of floated active silt and, hence, for microbe biomass concentration increase in the obtained foam product.

On the basis of pursued tests on lab sample of flotation apparatus, it's been established that it's possible to heighten microbe biomass concentration during movement of floated active silt in the channel with narrowing along foam product section. Meanwhile, decisive role in packing of the foam product of floated active silt is played by coalescence of foam gas bubbles. Speed growth of bubble coalescence process can be reached by the way of active silt contacting with hydrophobic surface as well as by the way of change of foam pH and other physical factors, for instance, vibration of foam layer of floated active silt or by the way of heating and air additional supply into the foam.

Flotation column experimental sample has been made and tested, the sample represents a vertical chamber of rectangular shape. Upper part of the column has pyramidal shape which ends with trumpet with foam gutter on its exterior side. Aeration-flotation chamber, which shape corresponds to the column shape, with plane bottom is located inside the column. This chamber height is 2/3 of the chamber height and its square correlates with the column square as 0.7:1.0. Five vertical jet aerators are fixed in the aeration-flotation chamber. Laminar thickeners are situated between walls of the chamber and the column.

Distinctive feature of column is a small height of working zone (2 m^2), possibility to work without supply of compressed air but with aeration by atmospheric air or any other gases as well with aeration of pulp and combining of aeration by air or by bubbles being distinguished from the solution.

Pressure-head apparatuses are used on biochemical productions for excessive active silt thickening. Active silt suspension from secondary decanters goes into an accepting capacity. Then, it's supplied by pump into a saturator where saturation with air proceeds, the air is sucked by an ejector. Intensification of air dissolution in saturator is reached by load in it of a cap made from Raschig tubes for to increase contact interface between liquid and gas phases.

Table 13.6. *Results of technological trials of pressure–head flotation skimmer on excessive silt thickening stage*

Day/sample number	Pressure in saturator, kgs/cm²	Rate (expenditure), m³/h, of			Active silt concentration, g/l		
		Initial suspension	Foam product	Cleared water	In initial suspension	In foam product	In cleared water
1	Active silt flotation in regular regime with silt saturation directly by air (the control)						
1/1	5.2	46	13	33	6.54	11.4	3.92
1/2	5.2	46	14	32	5.79	12.26	2.22
1/3	5.3	66	14	52	6.07	11.3	2.41
1/4	4.5	75	14	61	7.95	12.7	1.96
2	Active silt saturation with the use of working liquid						
2/1	5.4	72	15	57	5.25	11.6	4.12
2/2	5.4	72	15	57	4.62	7.62	3.00
2/3	5.4	68	15	53	6.27	7.18	5.37
2/3	5.4	68	15	53	5.89	8.26	4.96
2/4	5.4	68	15	53	6.56	12.2	3.45
3	Flotation with air supply into pipeline						
3/1	5.4	68	20	48	9.3	14.2	6.5
3/2	5.4	68	20	48	7.24	6.9	3.15
3/3	5.5	65	19	46	8.67	10.4	4.43

Active silt suspension, saturated by air via throttling device, is supplied into flotation reactors where gas is extracted in the form of tiny bubbles directly on active silt flakes. Being formed aggregates are separated from water as a result of surface-emergence in flotation separators. This scheme specificity – dilution of active silt suspension with under-silt water, being supplied consistently from the first flotation separator into the next one. Flotation process intensification is reached as a result of such dissolution. Pressure-head flotation apparatus, fixed on purification facilities of a biochemical production, provides for excessive active silt thickening till 94–96% humidity at gab with under-silt water till 500 mg/l of suspended substances.

Technological trial results are shown in Table 13.6. Data analysis shows that thickening degree of active silt is 6–8 relating to silt concentration in initial suspension.

Flotation thickening of silt can be made as at immediate saturation of silt by air as at use of working liquid. Ratio of volumes of the working liquid and the silt depends on initial concentration and properties of the silt. As a rule, the ratio is established experimentally, in 0.5–3:1. The working liquid or its mixture with the silt is saturated with air in a saturator. The air is supplied into saturator or a pressure-head pipeline with compressor or by the way of ejection into pressure-head pipeline.

Data, presented in Table 13.6, testify expediency and efficiency of flocculant usage and of additional foam concoction on the stage of active silt thickening.

13.4 USAGE OF MICROORGANISMS AS FLOCCULANTS

Usage of inorganic coagulants including iron and aluminum as usual in combination with polyacrylamide or other synthetic flocculants for wastewater purification and clearing of fine-dispersed suspensions isn't always effective. It's worth to denote also that dose increase of applied reagents for thickening process intensification leads to rise of costs. In this regard, attempts are made to use wastes of various productions qua reagents for thickening intensification of suspensions. The usage of microorganisms of excessive active silt, being formed at wastewater biochemical purification, represents special interest.

Table 13.7. Yeast influence on flotoconcentrate sedimentation speed

Yeast rate, kg/tn, of phosphorite	Sedimentation speed of phosphorite suspension, m/h	Solid content in drainage after suspension thickening, mg/l
0	5.1	270
0.300	7.2	122
0.660	11.1	30
1.200	11.8	28
1.500	12.1	28

Table 13.8. Influence of active silt microorganisms on flotoconcentrate sedimentation speed

Rate of active silt microorganisms, kg/tn	Sedimentation speed of phosphorite suspension, m/h	Solid content in drainage after suspension thickening, mg/l
0	5.1	270
0.500	6.9	98
0.750	9.5	45
0.900	11.8	35
1.200	12.1	30
1.500	12.9	30

Possible ways of excessive active silt utilization haven't been determined yet at present. The usage of active silt microorganisms qua reagents for the thickening of suspensions can appear one of the most promising methods of excessive silt utilization.

Taking into account a big practical significance of the question being considered, research has been held on influence of yeast and bacterial cultures on thickening process intensification of various types of phosphorite floto-concentrates.

Research methodology was in monitoring of sedimentation of suspensions of phosphorite flotoconcentrate of $-0.074-0.040$ class in cylinder of 0.5 l capacity and in definition of drainage turbidity after solid phase sedimentation. Active silt microorganisms were introduced into being thickened suspension with 8 g/l concentration. Yeast was supplied in the form of suspension of 12 g/l concentration. Content of solid in being thickened suspension — 15%. Experimental research results on microorganism influence on sedimentation rate are shown in Tables 13.7, 13.8 and on Fig. 13.5, 13.6.

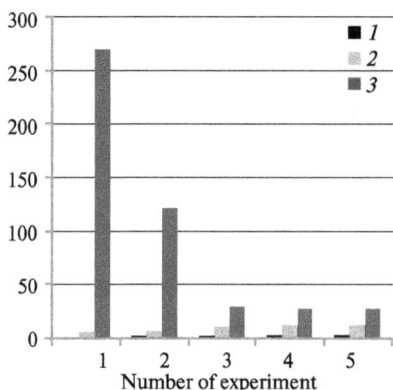

Fig. 13.5. Yeast influence on flotoconcentrate sedimentation speed:
1 — Yeast expenditure, kg/(tn of phosphorite);
2 — Sedimentation speed of phosphorite suspension, m/hr;
3 — Solid content in sink after suspension thickening, mg/l

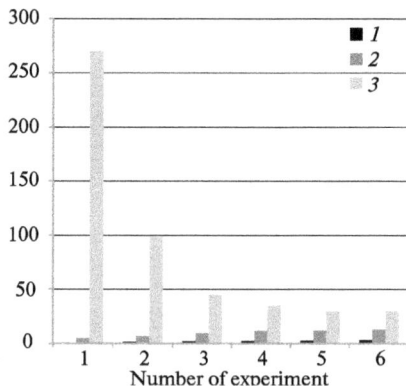

Fig. 13.6. Influence of active silt microorganisms on flotoconcentrate sedimentation speed:
1 — Yeast expenditure, kg/(tn of phosphorite);
2 — Sedimentation speed of phosphorite suspension, m/hr;
3 — Solid content in sink after suspension thickening, mg/l

Solid phase content in suspension, %	Rate of active silt microorganisms, mg/l	Speed gradient at mixing of suspension with microorganisms, c^{-1}	Drainage turbidity, mg/l
15	180	10	186
15	180	25	162
15	180	40	96
15	180	50	78
15	180	58	52
15	180	70	48
15	180	95	36
15	180	120	58
15	180	150	87

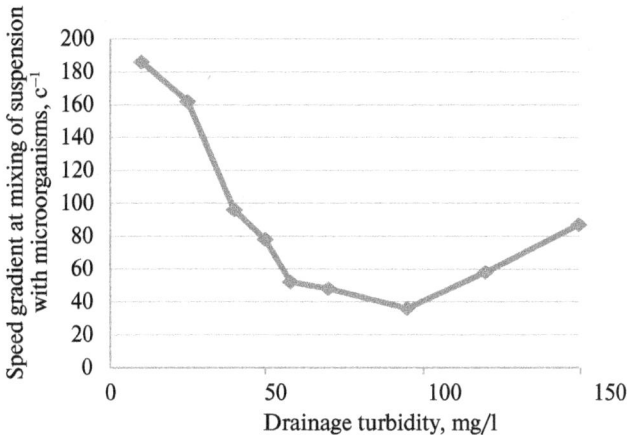

Fig. 13.7. Mixing intensity impact on turbidity of drainage
of being thickened flotoconcentrate phosphorite suspension
(at solid phase content in suspension of 15% and rate of active silt
microorganisms of 180 mg/l)

The obtained results show that sedimentation speed of phosphorite suspension flotoconcentrate increases as microorganisms in the suspension increase.

Treatment efficiency of being thickened with microorganisms suspension largely depends on optimal contacting conditions. In this regard, experimental research has been pursued to study the impact of blending intensity, characterized by speed gradient, on drainage turbidity of phosphorite suspension being thickened.

Research methodology was in the mixing of phosphorite suspension with active silt microorganisms in apparatus with propeller mixer. The speed gradient was determined by known methodology. The turbidity for drainage of being thickened suspension was measured after solid phase sedimentation. As a flocculant, we inputted active silt microorganisms. Solid content in suspension of phosphorite flotoconcentrate was constant and constituted 15%.

Experimental research results on mixing intensity impact on drainage turbidity are demonstrated in Table 13.9 and Fig. 13.7.

Experimental result analysis shows that speed gradient optimal value exists which at, drainage turbidity of being thickened suspension is minimal.

The obtained results might be considered using known provisions of gradient coagulation, playing main role at flake formation in fine-dispersed suspensions. Gradient coagulation process in the general form is described by equation:

$$n = n_0 \exp\left(-\frac{\psi}{\pi} C_0 G\tau\right),$$

where n — concentration of particles of solid phase suspension after time τ;

n_0 — initial concentration of particles of solid phase suspension;

C_0 — volume concentration of active silt microorganisms;

G-speed gradient.

Speed gradient, in its turn, can be calculated from the expression:

$$G = \sqrt{\frac{N}{V\eta}},$$

where N — wattage, spent on blending, W;

V — being mixed volume, m^3;

η — suspension dynamical viscosity.

In the case of use of apparatuses with mixers:

$$G = \sqrt{2\pi n \frac{M}{V\eta}},$$

where M — initial wattage, spent on blending, W;

n — mixer rotation frequency, s^{-1}.

Table 13.10. pH impact on flotation packing of active silt and its efficiency at further usage qua a flocculant for phosphorite concentrate thickening

Characteristics of active silt initial suspension		Active silt concentration in foam layer, formed as a result of pressure-head flotation (flotation time, hr) (g/l)	Dose of active silt, used qua a flocculant (g/l)	Concentration of solid phase in cleared liquid after separation of phosphorite concentrate suspension by defecating during 30 min (g/l)
Concentration (g/l)	pH			
4.5	6.9	24.2	0.25	0.64
4.5	4.0	38.6	0.25	0.58
4.5	2.0	49.3	0.25	0.34
4.5	1.5	50.1	0.25	0.31
4.5	1.0	50.1	0.25	0.30

Presented equation, describing flake formation process, doesn't take into account flake destruction process at the expense of speed large gradients appearing at mixing. Existing of gradient optimal values points at need to take flake destruction into consideration. Being limited with flake formation qualitive analysis with application of active silt qua flocculant, we denote that speed gradient at mixing of being cleared suspension with active silt shouldn't exceed $90-100$ s^{-1}.

Preliminary compression of active silt is very important at its usage as a flocculant. It's known that one of the simplest and most effective way of active silt compression before its application qua a reagent, and more precisely, qua a flocculant, is a pressure-head flotation. Together, active silt floatability depends on the series of factors, including also conditions of its cultivation, particularly, at large accumulation in active silt flakes of well soluble gases such as carbon dioxide, ammonia, the active silt is inclined to self-floatability. Formed thus foam layer is friable with not large concentration of the active silt. The usage of self-floatability of the active silt in combination with pressure-head flotation doesn't also allow to get compact foam layer with active silt high concentration.

Research, pursued by us earlier, has shown that one of the most effective ways of the process of active silt packing is its acidification. Meanwhile, decrease of pH is especially feasible at joint usage qua reagents for thickening of active silt phosphorite concentrate and of acid, for instance, phosphoric.

Mechanism investigation for pressure-head flotation of active silt with its preliminary acidification has demonstrated that at pH up to $1.5-2.0$, carbon dioxide solubility lowers drastically and it's extracted from liquid

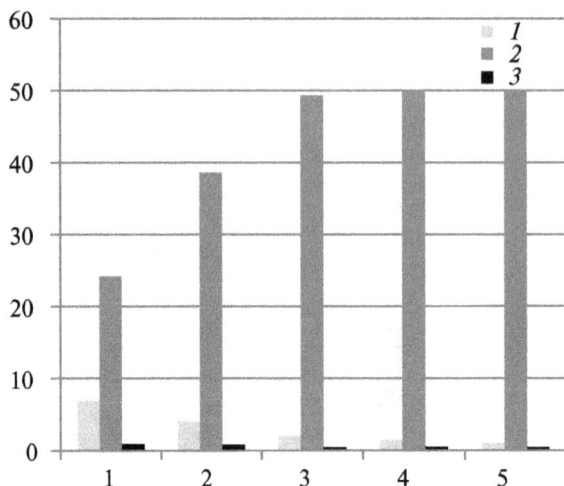

Fig. 13.8. pH impact on flotation packing of active silt and its efficiency at further usage qua a flocculant for phosphorite concentrate thickening (at concentration of active silt initial suspension 4.5 g/l and dose of active silt, used qua a flocculant, 0.25 g/l): *1* — pH; *2* — Active silt concentration in foam layer, being formed as a result of pressure-head flotation (flotation time, h) (g/l); *3* — Concentration of solid in defecated liquid phase after separation of phosphorite concentrate suspension by defecating during 30 min (g/l)

phase, including also from active silt flakes. Meanwhile, active silt flakes after exit from them of carbon dioxide become compact, and further pursing of pressure-head flotation leads to dense foam layer formation with active silt concentration in foam layer without preliminary acidification of the active silt being exposed to compression by flotation. Table 13.10 and Fig. 13.8 demonstrate evidences on both, pH impact on flotation compression of active silt or pH further impact on clearing efficiency of phosphorite concentrate suspension of −0.074 mm class with initial concentration of solid 5% of mass.

Data, presented in Table 13.10 and Fig. 13.8, testify that preliminary acidification of active silt till pH 1.5–2.0 drastically improves not just its flotation thickening but also sorption characteristics of the active silt that leads to vivid lowering of concentration of solid in liquid phase of being cleared fine-dispersed suspension of phosphorite concentrate.

Photographic research shows that extraction of carbon dioxide from flakes at pH 1.5–2.0 leads to that in liquid phase of active silt suspension at pressure-head flotation pursuing, air small bubbles of about 0.001–0.05 mm diameter and carbon dioxide bubbles of size up to 0.1–0.5 mm can exist. It leads to that complex (aggregate of fine-dispersed particles)-(air bubble)-(carbon dioxide bubble), possessing large lifting force, is formed, that, perhaps, results in growth of speed of active silt flotation and formation of dense foam layer with active silt high concentration.

14 FLOTATION PROCESSING OF WASTE PRODUCTS

14.1 FLOTATION PROCESSING OF A COAL ASH OF THERMOELECTRIC POWER STATIONS

One of the resources, that is actively mined and used in Russia, is a coal. Russia's share in the world stores of a coal constitutes about 11%. But mining and usage of a coal as well as of other mineral-raw resources create a series of ecological problems [41]. One of such problems, which is seriously faced today, is the problem of utilization of waste relating to coal mining and enrichment as well as to ash and slags formed at coal burning. Besides, usage of low-sort coals leads to quantity increase of mineral wastes formed as a result of their burning. Thus, slag- and ash-heaps are the sources of constant polluting of natural environment. At present, these are mainly the wastes that are used in no way, which are dropped in the form of dumps that results in the necessity to subtract big territories and in the expenditures on ash-slag wastes and on building of new territories.

That's why, from our point of view, the most perspective settlement of utilization of mining dumps and coal usage is a waste processing with the purpose to extract noble metals — gold and silver, and rare-earth metals — yttrium, scandium, lanthanum and others. Together, such approach not just settles ecological problems but can be also economically profitable because these metals are high-cost and in short supply.

Despite ecological and potential economic profit of noble and rare-earth metal extraction from ash- and slag-dumps, there are a series of problems. Undoubtfully, processing of ash-slag dumps with the purpose to extract noble and rear-earth metals is a complex and little-studied process. But, nevertheless, content of noble and rear-earth metals in dumps is rather little, hence the creation of technologies, that do not demand high capital and exploitational costs, is necessary.

If to consider precious metal extraction from ores, it's the most profitable economically and the least harmful ecologically the technology of biological leaching. Meanwhile, preliminary processing of ash-slag materials before bioleaching has a big importance. Our research has shown that the most perspective way is a preliminary flotation processing of ash-slag dumps of energetic enterprises [84].

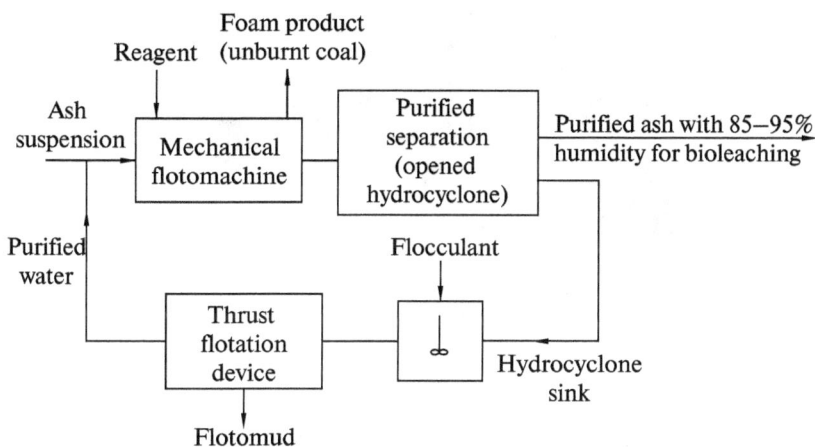

Fig. 14.1. Principal scheme of ash waste processing for subsequent
use in the process of bioleaching of rare-earth
and noble metals

Pursued in test-laboratorial conditions, experiments have allowed to develop improved way of processing of ash waste further usage as a raw in the process of leaching of rare-earth and noble metals. Together, it's allowed to choose floto-machine type and to define optimal regime of flotation treatment with water-use recursive system (Fig. 14.1).

The essence of being proposed approach is in the following. Ash wastes, being formed after coal burning, are mixed with a water with obtaining the fine-dispersed suspension wherein carbon containing reagent is inputted. Ash suspension flotation treatment is being held in mechanical floto-machine with flotation time of 15–18 min and air rate about 0.7–0.8 m^3/m^2min [8–10]. Chamber product in the form of purified ash suspension is supplied for separation into an open hydro-cyclone. Meanwhile, thickened product in the form of ash product with about 85–95% humidity is bent into special store space, and hydro-cyclone drainage is treated by cation-active flocculant, for instance, of praestol type and is directed afterwards for cleaning into pressure-head flotation device.

Qua a mechanical floto-machine — the developed by us flotation mechanical machine FCMP-0.15, which general view is shown on Fig. 14.2, is the most suitable for these purposes.

Sequence of ash treatment in floto-machine of such type includes ash homogenization with water and reagent (Fig. 14.3).

a

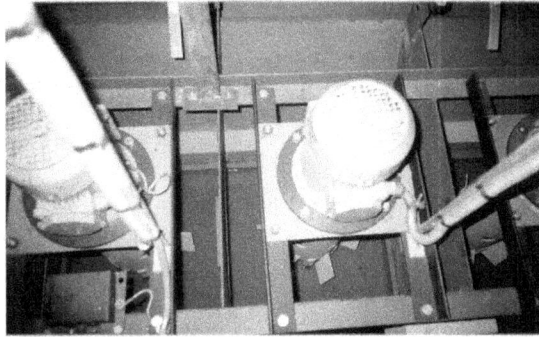

b

c

Fig. 14.2. Flotation mechanical machine FCMP-0.15:
a — machine general view; *b* — view on impellers from above;
c — sideways view on impellers.

Reagent

Water		
Ash	Ash homogenization with water and reagent	

Flotation separation of unburnt coal

Foam product in the form of unburnt coal

Flotation chamber product (purified ash) for bioleaching

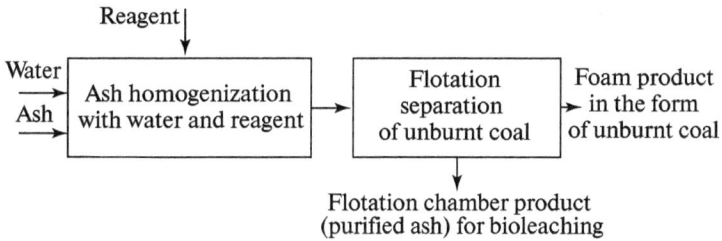

Fig. 14.3. Scheme of flotation treatment process

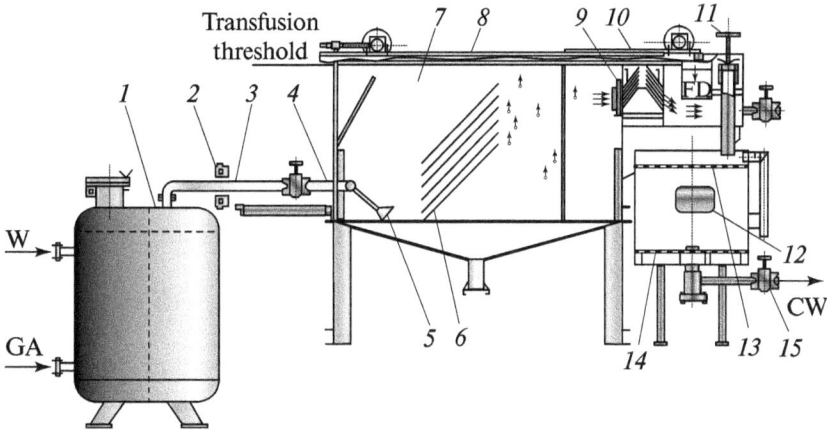

Transfusion threshold

W

GA

CW

Fig. 14.4. Pressure-head flotation device:
1 — pressure-head reservoir; *2* — magnetic treatment node;
3 — gas-water mixture supply pipeline; *4* — throttle device; *5* — trumpet;
6 — fine-zoned clearing block; *7* — flotochamber working space,
8 — foam removing device; *9* — fine-zoned device of microbubble catching;
10 — channel of output for co-coalesced micro-flotocomplexes;
11 — device of liquid level regulation; *12* — adsorption filter,
13, 14 — holding nets; *15* — latching device on output line of cleared-up liquid; W — dirty water supply; GA — gas-air flow supply; CW — cleared water output; FD — output of dirt in the form of foam product

Ash homogenization with water proceeds in the first chamber of multi-chamber (4-chamber) floto-machine. Together, impeller has linear rotation speed of impeller blade end of about 4—5 m/s that doesn't turn up to essential air sucking in flotochamber. The last brings that flotation process isn't observed. Reagent, for instance, kerosene, is added into the ash suspension at the exit of first chamber. After that, treated-up suspension comes into flotomachine second chamber.

Flotation processing of the ash suspension goes on in the second chamber of mechanical flotomachine. As a result, particles of unburnt coal are extracted into foam. Ash suspension, purified from unburnt coal, is outputted from flotomachine in the form of chamber product. Meanwhile, cleaned-up ash suspension contains no more than 5% of unburnt coal. Coal content in a foam product constitutes about 12–25%%.

After 15–40 minutes of the flotation treatment, unburnt coal content in ash doesn't exceed 5–12%%.

To accomplish the process of flotation processing of coal ash with recursive water-use system a pressure-head flotation device has been used (Fig. 14.4).

The use of such pressure-head flotation installation allows to get purified water with indicators on suspended substances of not more than 5–8 mg/l but on common carbon — 0.5–1.0 mg/l. Such values of denoted indicators allow to apply recursive water-use system to this process.

In a whole, technological process for flotational processing of coal ash is proposed to be embodied according to the scheme shown on Fig. 14.5. Individual elements of this scheme have been elaborated on test objects in industrial conditions.

Purified ash can be further used qua a raw in the processes of farther treatment by the way of bioleaching from it of noble and rare-earth metals.

Scheme of such treatment of ash depends on the processing tasks, the bodily content of initial raw and demands for being obtained products as well as depends on specificities of hydro-metallurgy of noble and earth-rare metals [85–86].

Fig. 14.6 demonstrates the principled scheme of bacterial leaching of rare-earth metals from ash-cindery materials of enterprises of energetics.

Technological scheme of bacterial leaching of rare-earth metals includes several major stages:
- preparatory operation;
- preparation and accumulation of bacteria;
- leaching of target metals;
- division of pulp and regeneration of working solution.

Preparatory operations include grinding (out-grinding) of concentrate till set size, removal of impurities there from and concentrating of target metals (minerals) on concentration tables or by flotation with possible further electromagnetic separation. Pursuing technology of preparatory operations is shown on Fig. 14.7.

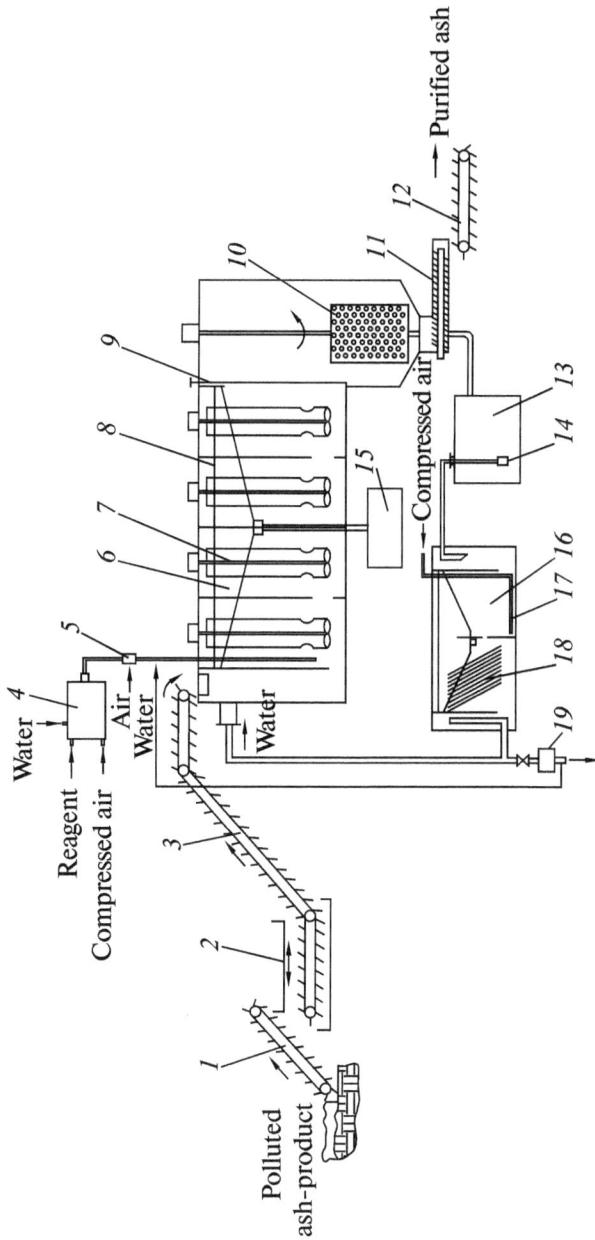

Polluted ash-product

Water
Reagent
Compressed air

Water
Air
Water

Water

Compressed air

Compressed air

Purified ash

Dropping into reservoir and then into 1st cell of floatation machine

a

1 2 3 4 5 6 7 8 9 10 11 12 13 14 15 16 17 18 19

257

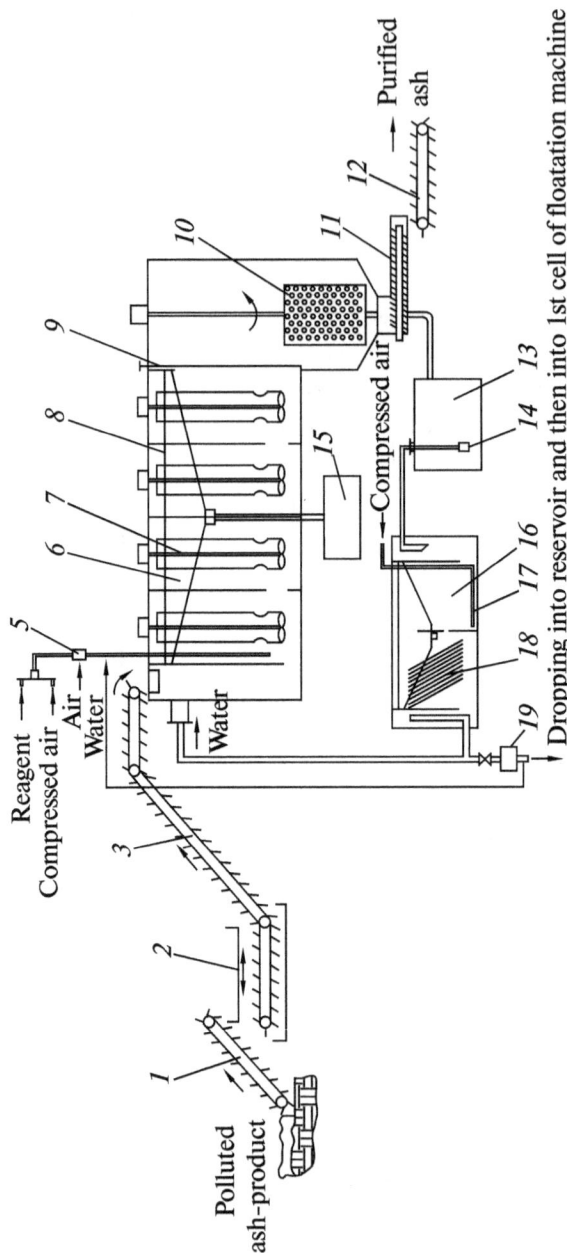

Fig. 14.5. Device scheme for ash waste processing (*a* — dry ash waste; *b* — ash waste from hydro-ash-removal system): *1* — device for mud and soil intake; *2* — lattice; *3* — transporter; *4* — node for reagent solution preparation; *5* — dosing node — ejector; *6* — mechanical flotomachine; *7* — aerators; *8* — foam gutter; *9* — device for regulation of level of suspensions; *10* — self-cleaning filter; *11* — screw machinery; *12* — transporter; *13* — intermediate reservoir; *14* — pump; *15* — foam collector; *16* — pressure-head flotation machine; *17* — aerators; *18* — fine-zoned block; *19* — polishing filter

Cell product of ash-slag materials

↓

Sampling

↓

Bacterial leaching
(biomass accumulation stage)

↓

Metal bacterial leaching

↓

Separation in decanter → Solution
for metal analysis

↓ ↑

Sediment dehydration ← Filtrate

↓

Sediment for analysis

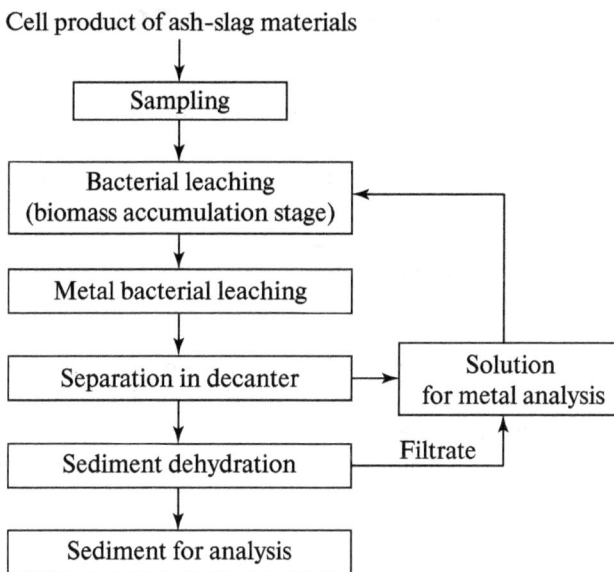

Fig. 14.6. Principled scheme of extraction technology
for rare-earth metals from ash-cindery materials

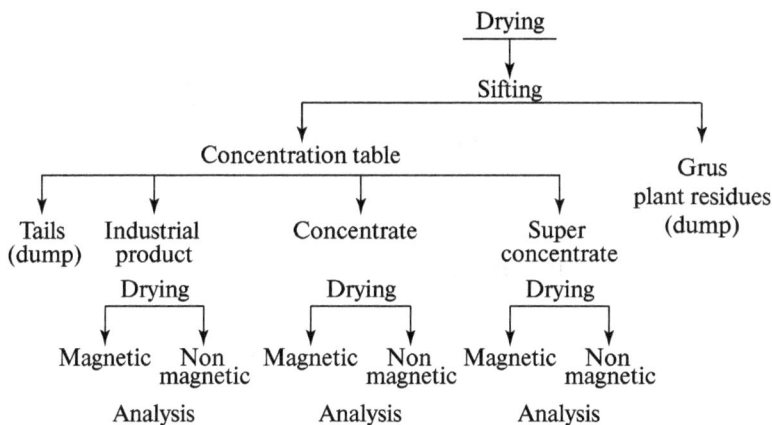

Drying

↓

Sifting

Concentration table   Grus
plant residues
(dump)

Tails   Industrial   Concentrate   Super
(dump)   product   concentrate

Drying   Drying   Drying

Magnetic   Non   Magnetic   Non   Magnetic   Non
magnetic   magnetic   magnetic

Analysis   Analysis   Analysis

Fig. 14.7. Technology of preparation of ash-cindery materials
for further bioleaching

Demonstrated schemes of coal ash treatment can be useful at the settlement of various practical ecological tasks.

14.2 RECOMMENDATION DEVELOPMENT ON UTILIZATION OF EFFLUENTS OF SLIPPER-BRAKE INDUSTRY

Different categories of wastewaters are formed in the production of slipper-brakes with the use of asbestine-technical items (ATI).

On ATI production, where trials were pursued, major sources of pollutions are presented by outgoing technological waters being formed in production workshops. These are aqua solutions of paste from chemicals, of penetrating-type compositions and so on. They're characterized by high content therein of such substances as alkali, glycerin, sulfur, zinc carbonate, technical carbon and many others.

To develop rational system of draining and to assess possibility of repeated usage of industrial wastewaters, their composition and draining regime are being studied. Physical-chemical indicators of wastewaters and regime of supply into canalization network of not just general drainage but of wastewaters from individual workshops and upon necessity from individual apparatuses are analyzed. Qualitative characteristics of industrial waters is important for choice of the method of their cleaning, the control on dropping and exploitation of purifying facilities as well as for resolving of questions on the possibility of their repeated usage, extraction and utilization of compounds polluting the water. Right collection of wastewaters is a requirement to obtain reliable and authentic results of the analysis.

Wastewater analysis of ATI production, pursued in "GosNIIsynthesbelok" Ltd, has shown their rather complex composition. Analysis of results on indicators of initial wastewaters are shown in Table 14.1.

Analysis of data, presented in Table 14.1, points at rather complex composition of wastewaters of ATI production.

One of the possible ways of industrial drainage utilization is the creation of water circulation on ATI production. Second usage of industrial wastewaters on enterprises in many cases is more advisable and cheaper than the preparation of fresh water for technological needs from water-supply sources.

Wastewaters often contain impurities which extraction is rather expensive but, nevertheless, is utterly necessary for provision of set defect action level. At the same, presence of these substances in the water doesn't harm the technological process. At repeated usage of wastewaters, as a rule, less degree of their purification is admitted than at the dropping into ponds. Besides, multiple usage of water, wherein limitedly soluble products of industry get in, allows to lessen their losses.

Table 14.1. Pollution content in drainages of ATI production

Indicators, mg/l	Initial drainages collected on September 1	Initial drainages collected on October 1
pH	7.2	7.4
Suspended substances	44	464
Ammonium nitrogen	44	47.6
Chlorides	34.5	183.6
Sulphates	108.8	140.7
Nitrates	n/a	n/a
Nitrites	n/a	n/a
Common phosphorus	2.0	2.8
Phosphates	n/a	n/a
Phenol	up to 1	up to 1
COD, mg O_2/l	58	716
Mineral oils, g/l	0.01	0.015
Nickel	0.1	0.1
Chrome	n/a	0.1
Copper	0.05	0.19
Zinc	0.4	16.6
Iron	0.5	9.3
Cadmium	0.01	0.01
Lead	0.05	0.1
Manganese	0.04	0.05
Fluorides	2.03	1.6
Arsenic, mercury	n/a	n/a

Requirements for repeatedly used water on ATI production haven't been worked out yet, and in this regard, it doesn't seem possible the industrial introduction of recursive water-supply at the moment.

To improve ecological situation on ATI production, it's needed to hold preliminary purification of drainages up to norms for dropping into city canalization (up to sanitary norms presented in Table 14.2).

One of the approaches to settle the problems of partial or full utilization of used tinctures is their application qua components for preparation of various technological liquids applied in drilling processes.

Table 14.2. Sanitary norms for content of impurities in drainages of ATI experimentative plant

Indicators	Sanitary norms, mg/l
pH	6.5–8.5
Suspended substances	257.0
Ammonium nitrogen	0.06
Chlorides	105.6
Sulphates	33.3
Nitrates	–
Nitrites	0.0062
Common phosphorus	–
Phosphates	–
Phenol	0.005
COD, mg O_2/l	750
Mineral oils, g/l	–
Nickel	–
Chrome	5.98
Copper	0.0004
Zinc	0.0026
Iron	0.0192
Cadmium	–
Lead	0.0165
Manganese	0.0071
Fluorides	0.0671

15 NEW DEVELOPMENTS OF PERSPECTIVE COMPOSITIONS OF FLOTATION MACHINES

15.1 NEW AERATION SYSTEMS

It's well-known that flotation process efficiency much depends on flotation machine constructions. As an example of new developments, let consider some flotomachines, worked out by the author and colleagues [53–55].

Flotation device of wastewater purification [53] has been developed which comprises node for preparation of water-air mixture consisting of pump with sucking line with entrance sleeves of wastewaters and air and with forcing line with exit nozzles; it also comprises flotochamber in the form of flotation column which external cylindric side on, bend sleeves of purified liquid and foam product are located, and inside — there is water semi burried wall which differs in that it's made as a reticular one and is plugged in to negative pole of continuous current supply; together, cylindrical electrode, plugged in to positive pole of continuous current supply, is additionally fixed between the column shell and reticular wall, coaxially to the column shell.

Besides, distance between the reticular wall and additionally fixed electrode is from 0.01 to 0.1 of their height; and fill of dispersal dielectric material is located between the reticular wall and additionally installed electrode, whereat the dispersal fill is made of sorption material and size of particles of the dispersal fill is from 1 to 5 mm.

Fig. 15.1 shows scheme of developed by us flotation installation of wastewater cleaning which includes node of aqua-air mixture consisting of Pump 3 with sucking line with entrance Sleeves of wastewaters 1 and air 2 and forcing Line 4 with exit Nozzle 8, and Flotochamber in the form of flotation column 7 where exterior cylindric side in, Sleeves of bend of purified water 13 and of foam product 5, being unloaded along inclined Bottom 6, are situated, and inside — there are mixing Chamber 9 and semi-submersible Wall 10. Meanwhile, the water semi reticular wall is plugged in to negative Pole of continuous current supply 14, whereat Electrode *11* is additionally fixed between the column shell and semi-submersible reticular Wall *10*. Distance S between semi-submersible reticular Wall *10* and Electrode *11* is from 0.01 to 0.1 of height H of denoted Electrodes *10* and *11*. In working space between Electrodes *10* and *11*, dispersal Fill *12* made of dielectric material, for instance, balls from ceramic glass which have size from *1* to *5* mm, is situated. There can be used qua such materials the porous spherical particles from organic glass which pore space is filled in with carbonaceous coal. The usage of porous spherical particles allows to heighten additionally wastewater purification efficiency at passing the space between Electrodes *10* and *11*. Additional activity of electrical field not just leads to additional effect but also increases efficiency of the usage of fill of dispersal porous dielectric material.

The usage of this flotation installation of wastewater purification allows to reach purification efficiency up to 96.2–98.7% by hydrophobic pollutions, in particular, by mineral oil products, oils and fats.

Work principle of flotation device for wastewater purification is in the following. Wastewaters are supplied via Sleeve *1*, and air — via Sleeve *2* which with the help of Pump *3*, are forced into Line *4*, and under excessive pressure, aqua-air mixture is passed via Nozzles *8*. As a result of passing through Nozzles *8*, aqua-air mixture is dispersed till tiny particles, in particular, sizes of air bubbles are of 0.1–0.5 mm, these bubbles at the contact with pollutions in mixing Chamber *9* form flotocomplexes particle-(small bubble) which emerge on surface in the working space of flotochamber, having the kind of flotation Column *7*. Flotocomplexes, lifting up to upper layer, form foam layer, which is removed along inclined Bottom *6* via Sleeve *5*. Being purified liquid is being bent via semi-submersible reticular Wall *10* and then via Layer of dispersal fill *12*. Together, water additional cleaning proceeds at the expense of activity of electric field between Electrodes *10* and *11*, plugged in to continuous current Supply *14*. Distance S between Electrodes *10* and *11* must constitute from 0.01 to 0.1 of height H of these electrodes. Dispersal fill particles must be in the limits of 1 to 5 mm and accomplished with dielectric material, possessing vivid sorption properties.

The use of proposed flotation installation allows to obtain wastewater purification efficiency up to 90–95%% at the lowering of energy rates in comparison with known technical settlement (prototype) by 1.5–2 times.

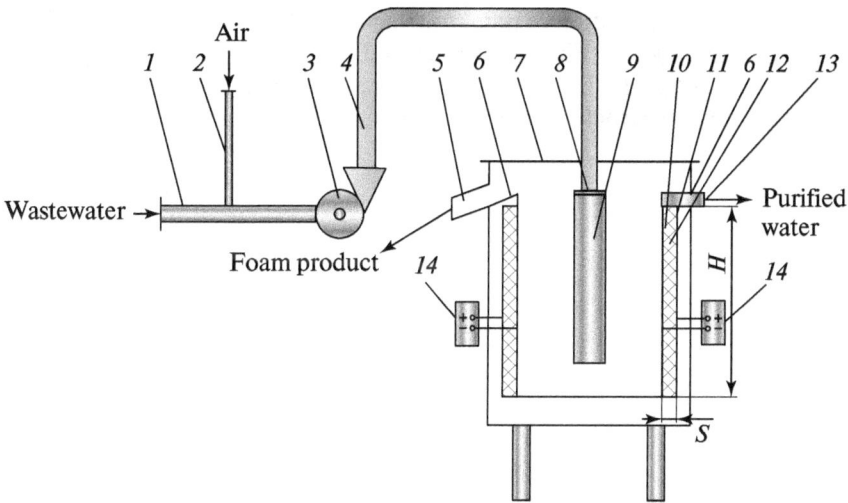

Fig. 15.1. Flotation installation scheme

Usage of above-considered installation is possible as at full capacity as in incomplete variant. In the case of application of aeration system only, used in this flotation device, different types of combined installations for water purification can be created on its basis.

For to purify wsstewaters with the use of reagent flotation, a flotation machine with conditioning chamber has been developed [54]. Whereat the development can be realized by two variants (Fig. 15.2–15.3). The proposed flotation machine for wastewater cleaning according to the first variant (Fig. 15.2) consists of Shell *1* divided by Walls *9* in chambers with fine-zoned defecating Block *10* in exit chamber. Inside Shell *1*, conditioning chamber is situated additionally which inside, Mixer *12* is fixed, having from *2* to *6* blades, with Drive *5* and liner Tube *6* with diameter D, having Windows *7* with diameter *d* (Fig. 15.4). Foam Gutter *4* with Sleeve *8* for foam bend is situated in the upper part of Shell *1*. Fone-zoned defecating Block *10* is in exit chamber of Shell *1*. Fine-zoned defecating Block *10* is located in exit chamber of Shell *1*. From the exterior side of Shell *1*, there are Sleeves of initial (dirty) water supply *2*, of reagents *3* and Sleeves of foam bend *8*, purified water *11*, as well as Sleeve for working liquid input *13*.

For to purify wastewaters with stable flakes (for instance, wastewater, polluted with mineral oils), the liner-tube can be accomplished in the form of cylinder (Fig. 15.2), and for cleaning of wastewaters with unstable flakes (for instance, purification of water from active silt) — in the form of conical diffusor with angle α (Fig. 15.3). Meanwhile, mixer of propeller type with inclination angle of blades β (Fig. 15.5) is established. Pointed approaches allow to preserve structure of being formed flakes, thus increasing purification degree.

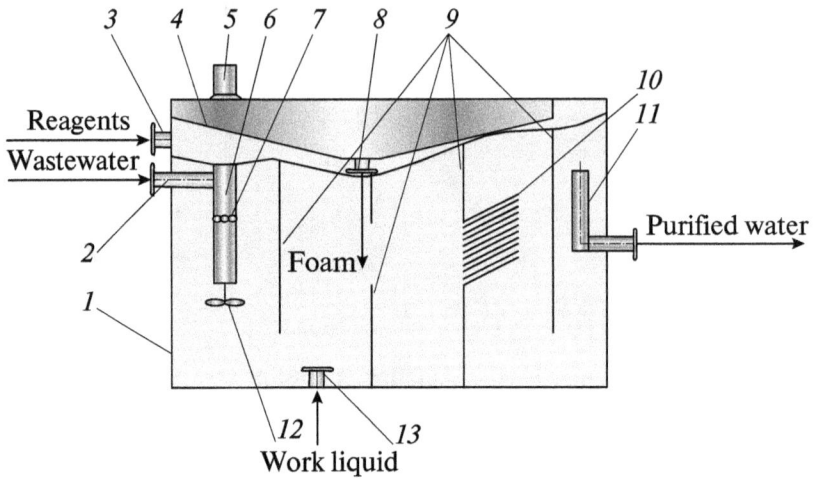

Fig. 15.2. Flotomachine scheme with conditioning chamber

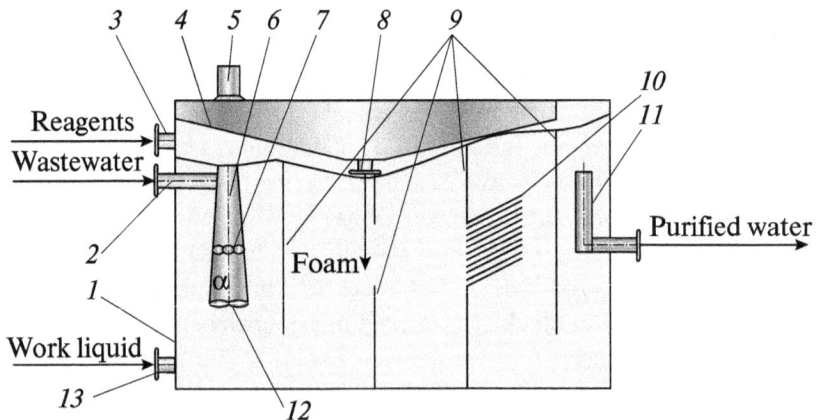

Fig. 15.3. Flotomachine with conditioning chamber, having mixer with liner-tube

$D = (1 - 5)d$

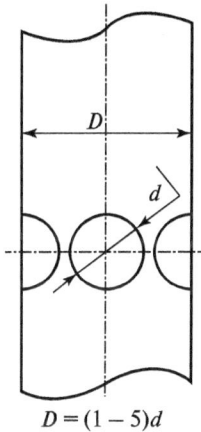

Fig. 15.4. Liner-tube scheme

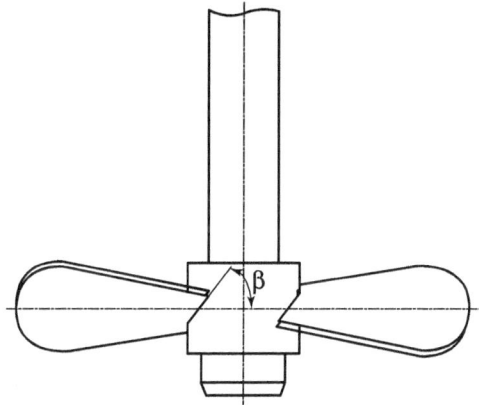

Fig. 15.5. Mixer scheme

Work principle of proposed flotation machine for wastewater purification is in the following. Initial (dirty) water via Sleeve 2 is supplied into Shell 1 of the flotation machine where via Sleeve 3, reagent solute and working liquid (gas-air mixture) are supplied via Sleeve 13.

Mixing of these flows proceeds at the expense of Mixer 12 rotation with the help of Drive 5. Efficiency of Mixer 12 action depends on blade number and their inclination angles to horizon blades β as well as depends on angle of conical diffusor α and area of Windows 7 in the liner-tube. Optimal parameters of angle α are from 5 to 45°, and diameter of the windows in the liner-tube — from 0.1 to 0.5 of the liner-tube diameter, and the number of windows — from 2 to 4. Together, inclination angle of mixer's blades constitutes from 15 to 75°.

As a result of the mixing of these flows, formation of flotocomplexes pollutions-(gas bubbles) goes on. The formed flotocomplexes surface-emerge, forming foam layer which's periodically poured into foam gutter wherefrom it's removed via Sleeve 8. The purified water goes through the system of shelves of fine-zoned defecating Block 10 and then via Sleeve 11, is outputted from Shell 1 of the flotation machine.

Area increase of windows in the liner-tube provides for rate increase of being pumped via them liquid that leads to increase of being seized from funnel, that's formed in the liner-tube, air as well as leads to the air finer dispersing. This is supported by pursued by us multiple experiments. Quantity growth of being sucked air and rise of its dispersal degree promote higher probability of capture of dirt particles by air bubbles. Obtained whereat spectrum of air bubbles at optimal parameters of the conditioning chamber leads to flotation time decrease and, respectively, to lowering of overall sizes of flotation machine. Embodiment of pointed phenomena in a whole helps purification efficiency by flotation approach.

Chosen intervals of values of areas for windows are substantiated by the way of experimental research holding. Together, window area choice, assessed by diameters, such as less than the area minimal value doesn't give rise of purification efficiency, and at the use of square values of windows as more than the maximal value, the reached effect doesn't increase.

Depending on wastewater kind and being used reagents the liner-tube might be fulfilled in the form of cylinder or conical diffusor with angle α in the frames of 5 to 45°, and together, blade number as 2 to 6 and optimal inclination angle of mixer blades amongst frames 15 to 75°, obtained experimentally, are chosen.

The denoted approaches allow to preserve being formed flake structure, thus, raising purification efficiency. Besides, reagent injection in this case appears more effective because reagent is being evenly spread across all chamber volume that promotes better formation and further preservation of flakes from dirt particles.

As a result of pursued by us series of experiments, it's been clarified that dirty (waste) water purification efficiency in the proposed flotation machine for wastewater purification is from 95 to 99% on hydrophobic pollutions, and at use of the installation-prototype — efficiency degree doesn't exceed 80.0%.

Thus, the offered flotation machine for wastewater cleaning permits to raise water cleaning efficiency approximately on 10–20%, permits to lower reagent rate at the expense of improvement of mixing on 20–25% as well as to decrease overall sizes, occupied by flotation machine, on 25–35%.

The flotation machine of original design for wastewater purification has been developed [55], it includes shell, divided by wall into flotation chamber and chamber for purified water with submersible pump, with, situated on its external side, sleeves for supply of dirty water, working liquid and air as well as for bend of purified water and foam product as well as it includes, fixed inside the shell, aeration device in the form of ejector and disperser, whereat the disperser is accomplished in the form of hollow body with corrugated surface, which corrugations perpendicular to, slot element is located; configuration and size of the element can be various, together, the hollow body might be in the form of cylinder or cone with obliquity angle of 10–60° with corrugations on both, interior and external side, and rods with semicylindrical shape are used as corrugations; the rods are fixed at mutual distance of 1...9 diameters of rods (Fig. 15.6). Together, the aeration device can be situated as on motionless as well as on mobile bases.

The proposed flotation machine for wastewater purification (Fig. 15.6) comprises Shell 1 with, arranged on its external side, Sleeves for dirty water supply 5, working liquid 2 and air 3 and purified water Bend 10 and foam Gutter 4 with Sleeve of foam product bend 6, Sleeve for purified water recycle 11, and inside the shell, there are aerating Device in the form of ejector 17, slot Element 16 and Disperser 1, semi-submersible Wall 7 with Net 14, and Chamber for cleaned water 9 with Regulator 8 of purified water bend and submersible Pump 12. Disperser (Fig. 15.7–15.14) is accomplished in the form of hollow Body 18 with corrugated surface with Corrugations in the form of semi-cylindrical rods 19, which perpendicular to, slot element is located. Meanwhile, the corrugations are accomplished as on internal (Fig. 15.14) as well as on external side of the hollow body (Fig. 15.7–15.10) which can be made in the form of cylinder (Fig. 15.7–15.8, Fig. 15.11–15.12) or truncated cone with obliquity angle of 10...60° (Fig. 15.9–15.10, Fig. 15.13–15.14).

The offered flotation machine working principle is in the following. Initial (dirty) water is supplied along Sleeve 5 (Fig. 15.6) into Disperser 15 area. The working liquid, which rate is regulated by Valve 13, is supplied by submersible Pump 12 and goes through Sleeve 2 inside Ejector 17 with high speed, whereat while passing through the ejector in water jet, the zone of lowered pressure occurs which is less than atmospheric one that leads to air sucking via Sleeve 3 and its blending with water jet. The obtained mixture via slot tangential Element 16 gets on corrugations of Disperser 15 in the form of semi-cylindrical rods (Fig. 15.7–15.14, stance 19), fixed on the hollow body surface (Fig. 15.7–15.14, stance 18). At the contact of aqua-air mixture and, first of all, gas phase with corrugations, the dispersal of the aqua-air mixture till tiny bubbles proceeds, together, intensive blending of the aqua-air mixture with the polluted water goes on that brings essential rise of water aeration degree and purification efficiency growth. Formed flotocomplexes surface-emerge forming foam which comes into foam Gutter 4 and is bent along Sleeve 6. Then the water passes through Net 14 and is bent along Sleeve 8 into purified water Chamber 9 wherefrom purified water major part is bent along Sleeve 10, and the other part — via Sleeve 11 for the use qua a working liquid.

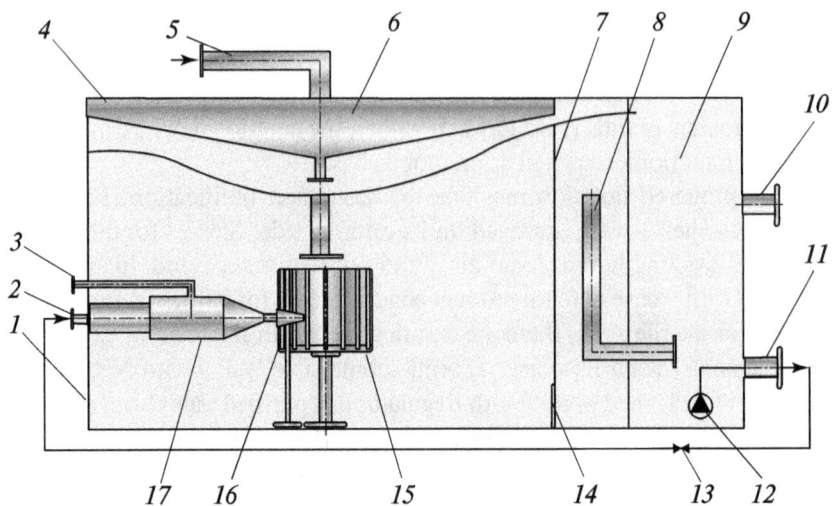

Fig. 15.6. Scheme of flotation machine with ejector and disperser

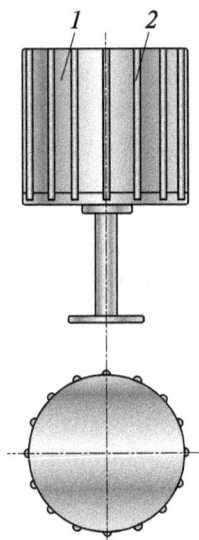

Fig. 15.7. Disperser of cylindric form with extrernal corrugations on motionless base

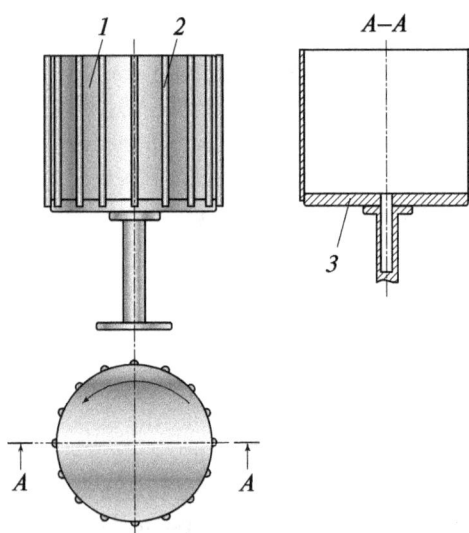

Fig. 15.8. Disperser of cylindric form with extrernal corrugations on mobile platform

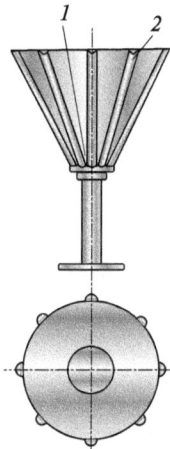

Fig. 15.9. Disperser of conical form with extrrernal corrugations on motionless platform

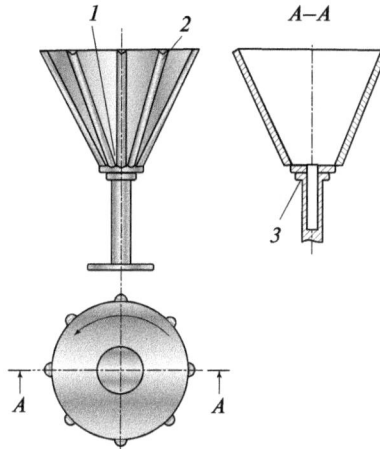

Fig. 15.10. Disperser of conical form with extrrernal corrugations on mobile platform

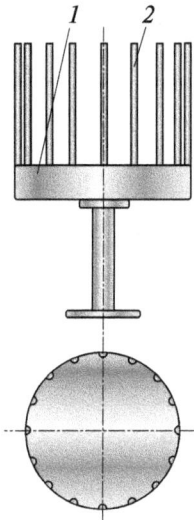

Fig. 15.11. Disperser of cylindrical form with internal corrugations on motionless platform

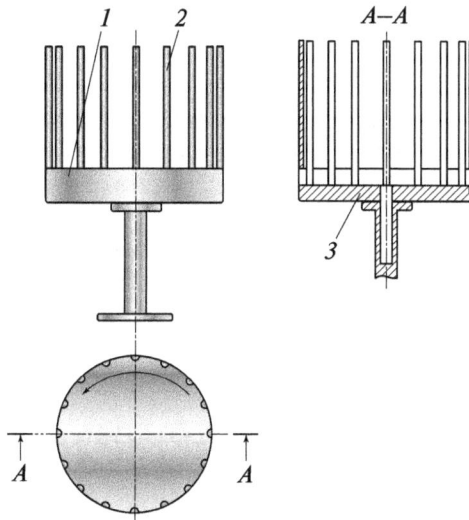

Fig. 15.12. Disperser of cylindrical form with internal corrugations on mobile platform

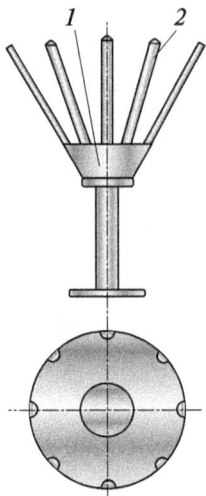

Fig. 15.13. Disperser of conical form with internal corrugations on motionless platform

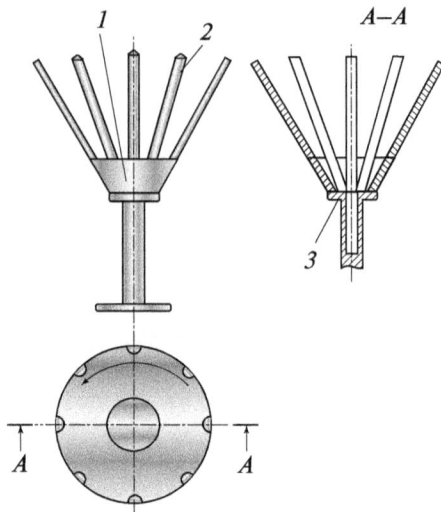

Fig. 15.14. Disperser of conical form with internal corrugations on mobile platform

The usage of novel technical settlements allows to raise purification efficiency on 15–20%% in comparison with known technical settlements at the pursuing of purification process of wastewaters, containing hydrophobic-hydrophilic pollutions.

The usage of the above-considered device is possible as in a full volume as well as in incomplete variant. In the case of application of just aeration system, used in this flotation device, one can create different types of combined installations for water purification on its basis.

15.2 FLOTATION HARVESTERS OF SPECIAL PURPOSE

Bioflotoharvester

Various flotoharvesters are known, they've been developed by us for purification of wastewaters wherein the processes of water purification and thickening of sedimentation proceed simultaneously [33–35].

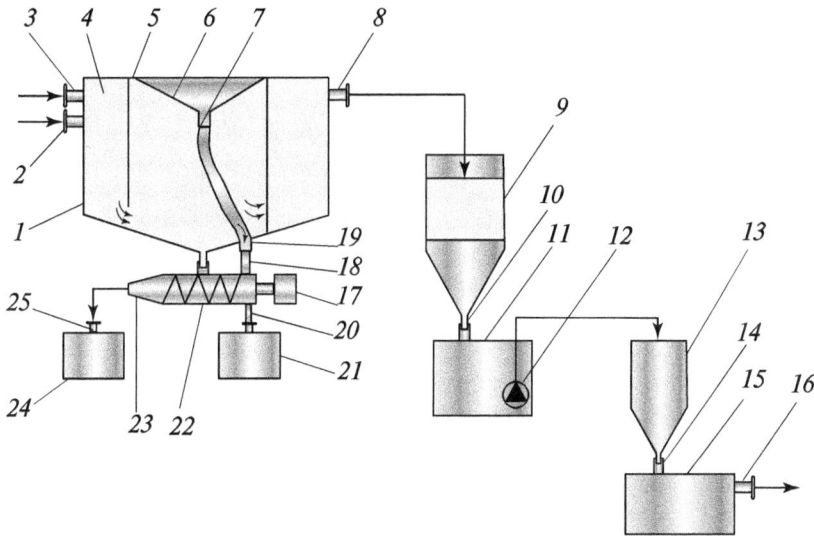

Fig. **15.15.** Bioflotoharvester scheme (Patent on useful model
N. 192973 "Bioflotoharvester", taken 06/18/2019, registered
10/08/2019. Applier and author Ksenofontov B.S.)

Together, in the pointed flotoharvesters, the most spread and productive
way of biological cleaning of wastewaters hasn't been realized that doesn't
allow to use widely this kind of water-purification technique for wastewater
purification.

In this regard, a device has been developed in the form of flotoharvester
[56]. Bioflotoharvester developed by us (Fig. 15.15) for wastewater purification comprises shell which inside, walls are located; on the external side, there
are fixed the sleeves, respectively, of supply for working liquid and wastewater,
caps, foam gutter with sleeve of foam product bend, sleeve of purified water,
foam product, bent via sleeve further along pipeline into worm thickener, having external drive, whereat to rise wastewater purification efficiency, the sleeve
of purified water is successively connected by pipeline with biofilter which
exit sleeve is attached to adsorption filter with bending into reservoir sleeve,
whereat adsorption filter diameter constitutes from 0.1 to 0.7 parts of biofilter
diameter, and adsorption filter height is from 0.8 to 1.9 parts of biofilter height.

The bioflotohavester (Fig. 15.15), being proposed, comprises Shell *1*
which inside, there are Walls *5*, and on the external side, there are fixed
Sleeves of supply, respectively, of working liquid *2* and wastewater *3*, Caps *4*,

foam Gutter *6* with Sleeve of foam product offtake *7*, Sleeve of purified water offtake *8*, connected by pipeline with Biofilter *9* which exit sleeve is connected with intermediate Reservoir *11* wherefrom being purified water with the help of Pump *12* goes into adsorption Filter *13* connected by Sleeve *14* with cleaned water Collector *15* having Sleeve *16* of purified water offtake on the external side. Foam product, bent via Sleeve *7* then along Pipeline *19* via Sleeve *18*, comes into worm Thickener *22* having Drive *17* as well as Sleeve *20* of cleared water output, being directed into Collector *21*. Thickened product through conical exit Sleeve *23* is bent into sediment Collector *24* via Sleeve *25*.

Working principle of being proposed bioflotoharvester is in the following. Wastewater via Sleeve *3* and working liquid are supplied into Shell *1* inwardly where combined flotation process of water preliminary cleaning proceeds. Further process of water polishing goes on in Biofilter *9* and then in adsorption Filter *13*. Then, purified water is supplied into clean water Collector *15* and is used as directed. Foam product, being formed at the accomplishment of combined flotation process of purification of water from foam Gutter *6*, is inputted via Sleeve *7* and Pipeline *19* into worm Thickener *22* wherein the foam product thickening proceeds, which after, thickened concentrate is supplied into sediment Collector *24*, and cleared water — into Collector *21*.

Quality of the obtained clean water, being selected from Collector *15*, meets requirements of water quality of ponds of fishery value, and sediment thickening degree makes it transportable with residual humidity from 65 to 80%, and at the use of known devices, the quality of purified water doesn't meet requirements of water quality for fishery value ponds by the majority of indicators.

The proposed bioflotoharvester can be used autonomically qua a local purifying device with square less than in the case of usage of facilities-analogs up to 1.5–2.0 times.

The proposed device in the form of bioflotoharvester is intended for wastewater cleaning from various pollutions.

Together, proposed bioflotoharvester can be used for cleaning of wastewaters of various branches of industry.

Chemoflotoharvester for wastewater purification

It's not been offered in afore-described flotoharvesters to use effective way of chemical processing of wastewaters with the application of strong oxidants, for instance, ozone, hydrogen peroxide and so on that eventually doesn't allow to reach high degree of wastewater purification.

In this regard, we've worked out chemoflotoharvester new design which provides for high degree of wastewater purification from pollutions [57].

Together, technical result of such design is the rise of efficiency of wastewater purification with the achievement of water quality allowing to realize the purified water dropping into ponds of industrial fishery purpose satisfying all being rationed indicators.

The set task and denoted technical result are achieved by that chemoflotoharvester for wastewater purification [57] includes shell which inside, walls are fixed, and on exterior side, sleeves of supply of working liquid and wastewater, respectively, as well as sleeve of offtake of preliminary purified water are made (Fig. 15.16). Meanwhile, chemoflotoharvester includes foam gutter with sleeve for foam product offtake along pipeline into worm thickener, having external drive, and in front of sleeve of wastewater supply, ejector for reagent input is fixed, whereat, chemoflotoharvester has cap, and sleeve of working liquid supply is linked additionally to ejector for ozone supply, besides, offtake sleeve of preliminary cleaned water is connected with the help of pipeline with chemical reactor which exit sleeve is linked to grainy filter with bending into reservoir sleeve, and together, after sleeve for offtake of preliminary purified water in front of chemical reactor, ejector for oxidant solute supply is fixed at equal distance between sleeve of preliminary purified water and chemical reactor.

The proposed chemoflotoharvester (Fig. 15.16) comprises Shell *1* which inside Walls *9* are located, and from external side, there are fixed Sleeve for working liquid supply *2* with located in front of it Ejector *4* with Sleeve for ozone supply *3*, Sleeve for wastewater supply *7* with fixed in front of it Ejector *5* with Sleeve *6* for delivery of a reagent, for instance, coagulant or flocculant solute, Caps *8*, foam Gutter *10* with Sleeve *11* for foam product offtake, Sleeve for purified water offtake *12* with fixed in the middle between it and chemical Reactor *15* Ejector *13* with Sleeve *14* for an oxidant supply. Whereat, chemical Reactor *15* is connected with the help of Sleeve *16* with intermediate Reservoir *17* wherefrom being purified water with the help of Pump *18* goes into grainy Filter *19*, linked by Sleeve *20* with purified water Collector *21*, having Sleeve *22* on external side for purified water takeoff. Foam product, being bent via Sleeve *11* further along Pipeline *25* via Sleeve *24*, is supplied into worm Thickener *28*, having Drive *23*, as well as into Sleeve *26* of output of cleaned water, being directed into Collector *27*. Thickened product via conical exit Sleeve *29* is bent into sediment Collector *31* via Sleeve *30*.

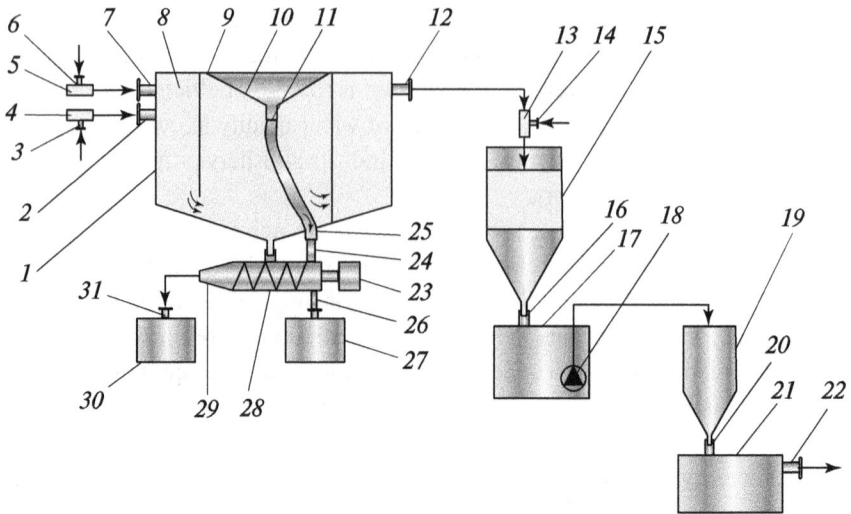

Fig. 15.16. Chemoflotoharvester scheme (Patent on useful model N. 194987 "Chemoflotoharvester", taken 08/09/2019, registered 01/10/2020. Applier and author Ksenofontov B.S.)

Working principle of being proposed chemoflotoharvester is in the following. Wastewater, preliminary treated with a reagent, inputted into wastewater via Sleeve 6 of Ejector 5, through Sleeve 7 and working liquid, mixed with ozone, supplied via Sleeve 3 of Ejector 4, and passing further through Sleeve 2, come inside Shell 1 where combined flotation process of water preliminary cleaning and disinfection goes on. Further process of water polishing proceeds in chemical Reactor 15 wherein being purified water is supplied via Ejector 13 with pumping through Sleeve 14 of this ejector of solute of oxidant, for instance, hydrogen peroxide. Besides, Ejector 13 is located in the center of the distance between Sleeve 12 of purified water offtake and chemical Reactor 15. Then purified water goes into clean water Collector 21 and is used by purpose. Foam product, being formed at accomplishment of combined flotation process of water purification from foam Gutter 10, is supplied via Sleeve 11 and Pipeline 25 into worm Thickener 28 wherein its thickening proceeds which after, thickened concentrate goes into sediment Collector 31 and cleaned waster — into Collector 27.

Quality of obtained clean water, being selected from Collector 21, meets requirements of water quality for dropping into ponds of fishing industry value of first and second category by all indicators being normalized, and thickening degree of sediment makes it transportable with residual humidity from 67 to 75%, together, with factually absence of nosey gases. At the use of known devices, purified water quality doesn't meet demands for water quality of ponds of fishing industry value by the majority of indicators and sediment represents a source of nosey gases.

Being proposed chemoflotoharvester can be used autonomically qua a local purifying installation with square less than in the case of using of facilities-analogs by up to 2.2–2.5 times.

Installation in the form of chemoflotoharvester is intended for wastewater purification from various pollutions.

Electroflotoharvester for wastewater purification

Being proposed device in the form of electroflotoharvester is intended for deep purification of wastewaters from various pollutions.

The purpose of the innovation is the creation of new design of electroflotoharvester that provides for wastewater high degree of purification from pollutions.

Technical result was embodiment of purified water dropping into ponds of fishing industry value of first and second categories. Set task and denoted technical result are reached by that electroflotoharvester for wastewater purification [58] includes flotoharvester which shell inside, walls and foam gutter with sleeve for foam product offtake into worm thickener, having external drive, are located, and on external side of flotoharvester shell, sleeves for supply of working liquid, wastewater and sleeve for purified water offtake are fixed, whereat, the flotoharvester contains cap, electroflotochamber is connected with the flotoharvester by means of sleeve for purified water offtake and pipeline, inside the flotochamber, there are perforated electrodes, located perpendicular to the direction of the flow of being purified wastewater with bipolar switching, from external side of the electroflotochamber, there are clean water offtake sleeve, sleeve of sediment offtake into worm thickener and foam gutter with sleeve for foam product offtake into worm thickener, and, together, distance between electrodes is from 0.01 to 0.09 to height of the electrodes, and perforation degree across live section is from 10 to 50%.

The proposed electroflotoharvester (Fig. 15.17) includes Shell *1* which inside, Walls *5* are situated, and from external side, there are Sleeves, respectively, for the supply of working liquid *2*, wastewater *3*, Caps *4*, foam Gutter *6* with

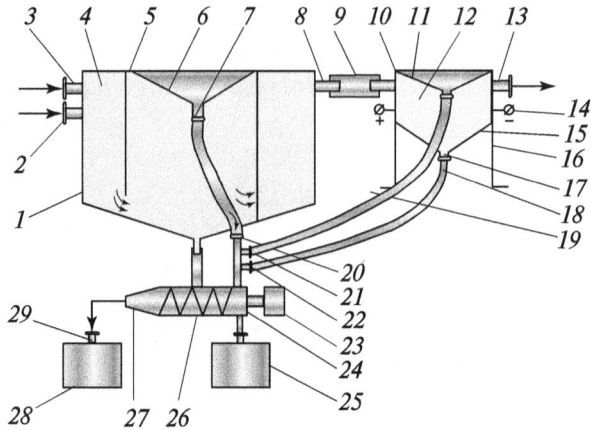

Fig. 15.17. Electroflotoharvester scheme
(Patent on useful model N. 194985 "Electroflotoharvester",
taken 08/22/2019, registered 01/10/2020.
Applier and author Ksenofontov B.S.)

Sleeve for foam product offtake *7*, Sleeve for purified water offtake *8*, connected to Pipeline *9* with Electroflotochamber *10* which external side on, there are fixed foam Gutter *11* for foam product offtake and its supply with the help of Pipeline *19* into entrance Sleeve *21* of worm Thickener *24*, exit Sleeve for clean water *13* and Sleeve for sediment offtake *17*, connected by Pipeline *18* with Sleeve *22* of sediment supply into worm Thickener *24*. Perforated Electrodes *12*, plugged in bipolarly to Clamps *14* of current source, are fixed inside the electroflotochamber. Together, distance between electrodes constitutes from 0.01 to 0.09 of height of the electrodes, and perforation degree, estimated by live section, might be equal from 10 to 50%. The given ratios were established and substantiated at the holding of experimental trials of the electroflotoharvester. The foam product, bent via Sleeve *7*, then goes by Pipeline *20* into worm Thickener *24*, having Screw *26* with Drive *23* as well as cleared water output into Collector *25*. Thickened product via conical exit Sleeve *27* is bent into sediment Collector *28* via Sleeve *29*.

Working principle of the proposed electroflotoharvetser is in the following. Wastewater via Sleeve *3*, and working liquid via Sleeve *2* are supplied inside Shell *1* where combined flotation process of water preliminary purification proceeds. Further process of water polishing goes on in Electroflotochamber *10* and wherefrom clean water is outputted via Sleeve *13* and is used by purpose.

Foam product, being formed at realization of combined flotation process of water purification, from foam Gutter 6 goes via Sleeve 7 and Pipeline 20 into worm Thickener 24 wherein there are supplied the foam product of its own accord and sediment from Electroflochamber 10 via Sleeves, respectively, 21 and 22. Thickening of mixture of the foam products and sediment proceeds in worm Thickener 24 which after, thickened concentrate is supplied into waste Collector 28, and cleared water — into Collector 25.

Quality of the obtained clean water, being bent from Sleeve 13, meets requirements of water quality for dropping into ponds of fishery industry value, and thickening degree of the mixture of foam products and sediment makes the mixture transportable with residual humidity from 65 to 75%. At the use of known devices, the quality of purified water doesn't meet requirements for water quality of ponds of fishery industry value by the majority of indicators, and the sediment's needed to be additionally dehydrated to make it transportable.

The proposed electroflotoharvester might be applied autonomously qua a local purifying installation with square less than in the case of the use of known facilities-analogs by up to 1.5–2.5 times.

It's worth to note that device in the form of electroflotoharvester is intended for wastewater purification from various pollutions including dissolved organic substances and particles of colloidal sizes.

Deminoflotoharvester for wastewater purification

Being proposed device in the form of deminoflotoharvester is designated for wastewater purification from various pollutions, including dissolved salts.

Whereat, the necessary degree of wastewater purification by salts hasn't been reached in known flotoharvesters. Development task is to create new design of deminoflotoharvester, providing for high degree of wastewater purification from pollutions, including salts.

Technical result is efficiency increase for wastewater purification with the achievement of water quality, allowing to embody purified water dropping into ponds of fishing industry value of first and second categories.

The set task and denoted technical result are reached by that deminoflotoharvester for wastewater purification [59] includes shell which inside, walls are located, and on external side, sleeves are installed, respectively, for the supply of working liquid and wastewater, caps, foam gutter with sleeve for foam product offtake, sleeve for offtake of purified water, foam product, being bent via sleeve further along pipeline into worm thickener, having external drive, differing in that with the purpose to raise efficiency of wastewater purifica-

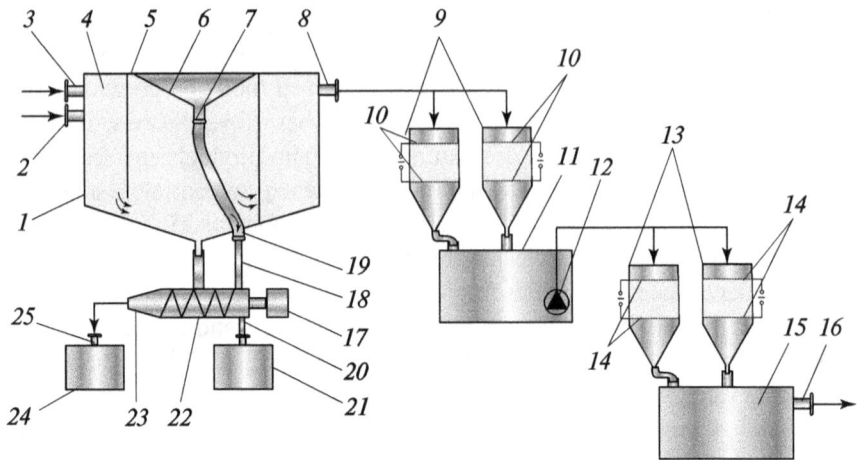

Fig. 15.18. Deminoflotoharvester scheme (Patent on useful model N. 194986 "Deminoflotoharvester", taken 09/16/2019, registered 01/10/2020. Applier and author Ksenofontov B.S.)

tion the purified water sleeve is connected by pipeline successively with cation- and anion-exchange filters, whereat ion-exchange cap therein is between electrodes, plugged in to alternating current source and whereat, reticular electrodes with quadratic cells are used, the cell size is from 3 to 10 mm.

The proposed deminoflotoharvester (Fig. 15.18) includes Shell 1 which inside, Walls 5 are situated, and from the external side, there are Sleeves, for the supply, respectively, of working liquid 2 and wastewater 2, Caps 4, foam Gutter 6 with Sleeve for foam product offtake 7, Sleeve for purified water offtake 8, linked by pipeline with cation-exchange Filters 9 wherein cation-exchange cap between reticular Electrodes 10 with quadratic cells with quadrate side from 30 to 10 mm, whereat which exit sleeves are connected with intermediate Reservoir 11 wherefrom being purified water with the help of Pump 12 is supplied into anion-exchange Filters 13 with located therein reticular Electrodes 14 which between, there is anion-exchange cap with quadratic cells with quadrate side of 3–10 mm, the cells are linked with purified water Collector 15, having Sleeve of purified water offtake 16 on external side. Data on the cell shape and size were established and substantiated during pursuing of model trials of deminoflotoharvester. Foam product, being bent via Sleeve 7, then along Pipeline 19 through Sleeve 18 is supplied into worm Thickener 22, having Drive 17, as well as Sleeve 20 of cleared water, directed into Collector *21*. Thickened product via exit conical Sleeve *23* is bent into sediment Collector *24* via Sleeve *25*.

Working principle of being proposed deminoflotoharvester is in the following. Wastewater though Sleeve *3* and working liquid through Sleeve *2* are supplied inside Shell *1* where combined flotation process of water preliminary cleaning proceeds. Further process of water polishing from dissolved salts, namely, from cations, goes on in cation-exchange Filter *9*, in which connection, one filter is in the working regime, while the second one — in regeneration regime, and then anion extraction proceeds from anions in anion-exchange Filter *13*. Meanwhile, one filter from anion-exchange group is exploited, while second one — is regenerated. Intensification of ion-exchange processes is achieved by apposition of alternating electric current on ion-exchange cap: on cation-exchange cap — in cation-exchange filters and, correspondingly, on anion-exchange one — in anion-exchange ones by means of providing of electric supply on reticular electrodes with quadratic cells and the size of quadrate side is from 3 to 10 mm. Choice and substantiation of the denoted measurement has been accomplished on the basis of experimental research.

Further, purified, including from salts, water is supplied into clean water Collector *15* and's used by purpose.

Foam product, being formed at the realization of combined flotation process of water purification, is supplied from foam Gutter *6* through Sleeve *7* and Pipeline *19* into worm Thickener *22* wherein its thickening proceeds which after, thickened concentrate goes into sediment Collector *24*, and cleaned water — into Collector *21*.

Quality of obtained water being selected from Collector *15*, meets all the requirements of water quality, including ones by dissolved salts, for dropping into ponds of fishing industry value of the first and second categories, and sediment thickening degree makes it transportable with residual humidity from 65 to 80%%, and at the use of known devices, the cleaned water quality doesn't meet the demands for water quality of ponds of fishing industry value by series of indicators and sediment must be additionally dehydrated for follow-through till transportable state.

The proposed deminoflotoharvester can be autonomously used qua a local purification device with space of less than in the case of the usage of facilities-analogs by up to 1.5–2.0 times.

The device in the form of deminoflotoharvester is intended for processes of both, wastewaters from various pollutions or natural waters, being taken from springs with various salt-content.

Silt-flotoharvester

Development task was in the creation of new design of silt-flotoharvester, providing for high degree of water purification from dirt to the point of water quality with the possibility of dropping into ponds of fishing industrial value, and the obtaining of sediment with diminished residual moisture content.

Technical result is the increase of water purification efficiency with the possibility of dropping into ponds of fishery industrial value and the obtaining of sediment with diminished residual moisture content.

The set task and denoted technical result are reached by that silt-flotoharvester for water purification [60] includes shell which inside, walls are fixed, and from external side, sleeves for supply, respectively, of working liquid and being purified water, caps, foam gutter with sleeve for foam product offtake, sleeve for offtake of purified water and foam product, the latter is bent through sleeve and further along pipeline into worm thickener, having external drive and linked by pipelines with collectors of thickened sediment and of cleared liquid, whereat for to rise water purification efficiency and to obtain sediment with diminished residual moisture content, the sleeve of purified water is connected by pipeline successively with bioreactor which exit sleeve is linked to intermediate reservoir, having loading pump, and then successively to membrane apparatus having offtakes into foam gutter and purified water container, whereat cleared water collector is connected by pipeline with node of working liquid preparation, the node is linked by pipeline to sleeve of working liquid and, together, the node of working liquid preparation is at the distance from collector of cleared liquid as 1/7−1/3 of distance between cleared liquid collector and sleeve for working liquid supply.

The proposed silt-flotoharvester (Fig. 15.19) comprises Shell *1*, wherein Walls *5* are located, and from external side, there are Sleeves for supply, respectively, of working liquid *2* and wastewater *3*, Caps *4*, foam Gutter *6* with Sleeve for foam product offtake *7*, Sleeve of purified water offtake *8* connected by pipeline with Bioreactor *9* which exit sleeve is connected with membrane Apparatus *12*, containing Membrane *11* inside, being purified water from the apparatus via exit Sleeve *13* by pipeline is supplied via Sleeve *14* into clean water Reservoir *15* wherefrom being purified water is outputted via Sleeve *16* with further usage by purpose. Foam product, being bent through Sleeve *7* and further along Pipeline *19*, as well as concentrate from the membrane apparatus via Sleeve *18* are supplied into worm Thickener *22*, having Drive *17*, as well as Sleeve *20* of cleared liquid, being directed into Collector *21*. Thickened product via conical exit Sleeve *23* is bent into sediment Collector *24* via Sleeve *25*. Meanwhile, cleared water Collector *21* is linked by pipeline to

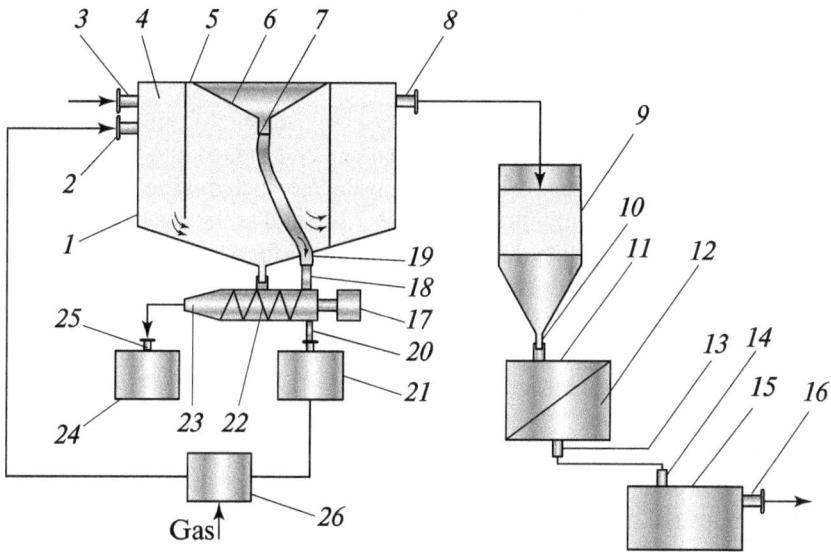

Fig. 15.19. Silt-flotoharvester scheme (Patent on useful model N. 195481 "Silt-flotoharvester", taken 10/25/2019, registered 01/29/2020. Applier and author Ksenofontov B.S.)

Node of working liquid preparation *26* which's connected by pipeline with Sleeve *2* of working liquid, together, the node of working liquid preparation is at the distance from the cleared liquid collector as 1/7–1/3 of distance between the cleared liquid collector and the sleeve for working liquid supply.

Working principle of being proposed silt-flotoharvester is in the following. Being purified water via Sleeve *3*, and working liquid via Sleeve *2* are supplied inside Shell *1* where combined flotation process of preliminary water purification proceeds. Further process of water polishing goes on in Bioreactor *9* and in membrane Apparatus *12*. Then purified water goes into clean water Collector *15* and's used by purpose. Foam product, being formed at realization of combined flotation process of water purification, is given from foam Gutter *6* via Sleeve *7* and Pipeline *19* into worm Thickener *22* wherein the product thickening goes on, which after, thickened concentrate is supplied into sediment Collector *24*, and cleared liquid — into Collector *21*. Together, Node of working liquid preparation *26* is at the distance from working liquid Collector *21* as 1/7–1/3 of distance between cleared liquid Collector *21* and Sleeve for working liquid supply *2*. These parameters have been checked at trials of test device.

Gas, for instance, air, together with liquid is supplied into Node of working liquid preparation *26* that leads to the formation of working liquid in the form of mixture gas-liquid.

Quality of being obtained clean water being selected from Collector *15* meets requirements of quality of water that can be dropped into pond of fishery industry value, and sediment thickening degree, including active silt, leads to rather low residual humidity from 62 to 73%, and at the use of known devices, purified water quality doesn't meet demands for quality of water, which can be dropped into ponds of fishery industry value, and residual moisture of sediment, including active silt, exceeds 80%.

The proposed silt-flotoharvester can be used autonomously qua local purification device of square less than in the case of the usage of facilities-analogs by up to 2.0–3.5 times.

Device in the form of silt-flotoharvester is intended for water purification form various pollutions till water quality, allowing to drop water into ponds of fishery industrial purpose, and for to obtain sediment, including active silt, in thickened state.

Soil-flotoharvester for water and soil purification

Task of this novel development is a creation of new construct of soil-flotoharvester, providing for high degree of cleaning of soils and water from dirt. Technical result is an increase of productivity of soil and water purification.

Set task and denoted technical result are achieved by that there has been offered soil-flotoharvester for water and soil cleaning [61] which comprises shell wherein walls are fixed, and from external side, there are sleeves for supply, respectively, of working liquid, being purified water and of soil, caps, foam gutter with sleeve for foam product bend, sleeve for bend of purified water, foam product, having external drive, whereat between walls, there are blenders with drives, fixed on the shell cap, and together, purified water sleeve is successively connected by pipeline with contact coagulation node, represented in the form of grainy filter with reagent supply device, which exit sleeve is linked to intermediate reservoir with loading pump and further successively to polishing filter and reservoir for water disinfection. Meanwhile, reagent supply includes pump — a dispenser of plunger type, and polishing filter fill comprises activated carbon and shredded anthracite in ratio, respectively, from 5:1 to 1:3. Fig. 15.20 represents soil-flotoharvester scheme.

Being proposed soil-flotoharvester (Fig. 15.20) includes Shell *1* which inside, there are fixed Sleeves for supply, respectively, of working liquid *2*, wastewater *3*, Cap *4*, blender Drives *5, 6, 7, 8*, foam Gutter *9* with bend of foam product along Pipeline *39*, connected via Sleeve *38* with Collector *37*,

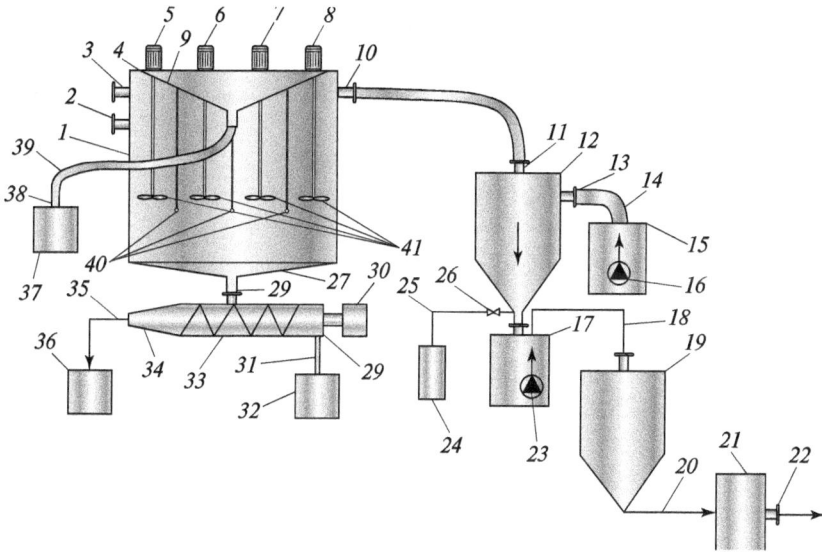

Fig. 15.20. Soil-flotoharvester scheme (Patent on useful model N. 195504 "Soil-flotoharvester", taken 11/05/2019, registered 01/29/2020. Applier and author Ksenofontov B.S.)

Sleeve for purified water bend *10*, connected by Pipeline *11* with grainy Filter *12*, and then, via Sleeve *13* by Pipeline *14* with Collector *15* with being inside it Pump-Dispenser *16*, for instance, of plunger type. Exit from grainy filter is linked to intermediate Collector *17* as well as connected by Pipeline *25*, whereon regulating Valve *26* is fixed, with Reservoir *24* for washing water release. Together, intermediate Collector *17* is hydraulically successively connected with the help of Pump *23* and Pipeline *18* with polishing Filter *19* and then with the help of Pipeline *20* with water disinfection Reservoir *21* having exit Sleeve *22*.

Polishing Filter *19* fill includes activated carbon and shredded anthracite in ratio, respectively, from 5:1 to 1:3.

Shell *1* conical Part *27* is connected with Pipeline *28*, having worm Thickener *29*, which has inside in its turn Screw *33* with Drive *30*. Together, on external side of worm Thickener *29*, there are on its cylindrical part Pipeline for cleared water bend *31*, connected with Collector *32*, and in its conical part Pipeline *35* for purified soil bend into Collector *36*.

Working principle of proposed soil-flotoharvester is in the following. Being purified water or soil aqua suspension via Sleeve *3*, and working liquid via Sleeve

2 are supplied inwards Shell *1* where combined flotation process of water preliminary cleaning proceeds. Further process of water polishing goes on in grainy Filter *12* wherein fill, for instance, from (Russian) sorbent AS (aluminum silicate) is located, and wherein coagulant solution, for instance, of vitriolic aluminum is supplied to. Then preliminary purified water goes into intermediate Collector *17* wherefrom with the help of Pump *23* by Pipeline *18*, it's supplied into polishing Filter *19* with fill from activated carbon and shredded anthracite in ratio from 5:1 to 1:3, respectively. Particular ratio choice depends on pollution kind in water and soil. After polishing Filter *19*, water by Pipeline *20* comes into Reservoir *21* of disinfection with application of natrium hypochlorite and then water is outputted from it via Sleeve *22* and is ready for use in various purposes.

Quality of obtained clean water, being selected from Collector *15*, meets requirements of water quality which can be dropped into ponds of fishery industry purpose, and soil purification degree allows to use it in a city planting of greenery.

Soil-flotoharvester can be used for purification of soil-grounds not only from mineral oils but also from heavy metals.

Aquaflotoharvester

Useful model "Aquaflotoharvester for water purification" task is a development of new design of aquaflotoharvester providing for water cleaning high degree from pollutions to the drinking point.

The set task and denoted technical result are achieved by that aquaflotoharvester [62] for water purification includes shell wherein walls are located, and from external side, there are sleeves for supply of, respectively, working liquid and being purified water, caps, foam gutter with sleeve for foam product bend, sleeve for purified water bend, which is connected with filter and then with pure water container with immersible pump therein for offtake of purified water, foam product which is bent further by pipeline into worm thickener having external drive, whereat after sleeve for foam product bend, before worm thickener, methane tank is fixed additionally which upper gas pipeline is connected with saturator for biogas solution preparation, and clean water reservoir is linked to saturators for preparation of working liquids, respectively, with well-soluble and poorly-soluble gas in purified water gas, meanwhile, working spaces of saturators for preparation of working liquids, respectively, with well-soluble gas and poorly-soluble one in purified water are related as from 1:1 to 1:10, accordingly.

Fig. 15.21 shows aquaflotoharvester scheme.

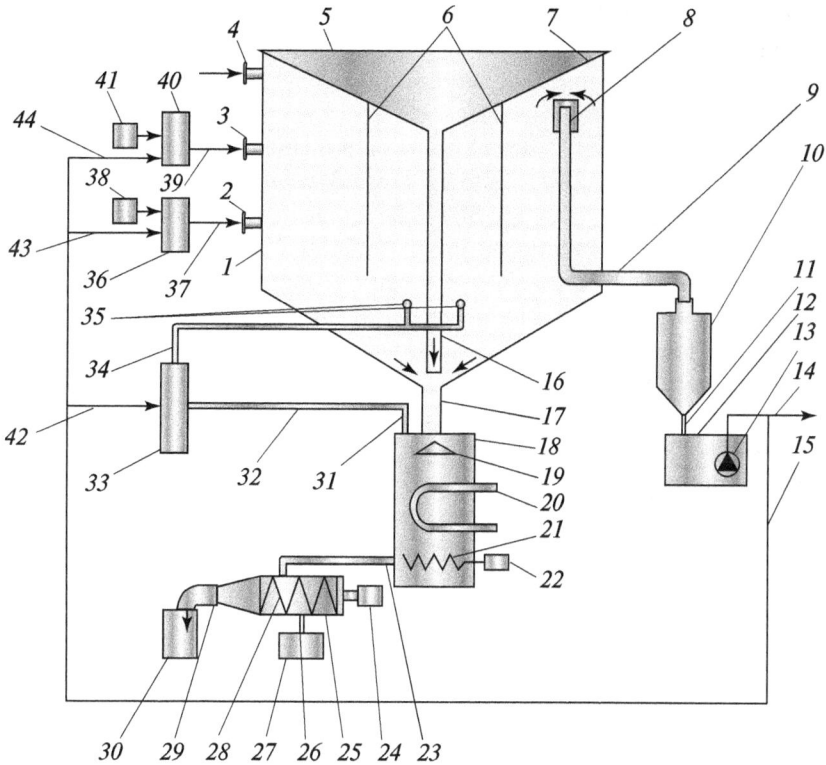

Fig. 15.21. Aquaflotoharvester scheme

Being proposed aquaflotoharvester (Fig. 15.21) includes Shell *1* which inside, there are Walls *6*, and from external side, there are fixed Sleeves for the supply of working liquids, respectively, with well-soluble gas by Pipeline *37* through Sleeve *2*, for instance, carbon dioxide being supplied from Balloon *38* and the gas is being prepared in Saturator *36* with supply into the saturator of purified water along Pipeline *43*, as well as with working liquid with poorly-soluble gas along Pipeline *39* via Sleeve *3*, for instance, with air, being prepared in Saturator *40* wherein there are supplied purified water via Pipeline *44* and compressed air from Compressor *41*, the aquaflotoharvester also includes Sleeve for dirty (waste) water *4*, Caps *5*, foam Gutter *7* with Sleeve for foam product offtake *16*, Device for purified water output *8* connected by Pipeline *9* with Filter *10* which exit Sleeve *11* is connected with intermediate Reservoir *12* wherefrom being purified water with the help of Pump *13* is supplied into Pipeline *14* wherefrom it's partially outputted by purpose and partially —

287

along Pipeline *15* which's linked to Pipelines *43* and *44* with Saturators *36* and *40*, respectively.

Foam product, being bent via Sleeve *17*, is supplied into methane Tank *18* wherein there're fixed Divider *19*, Serpentine *20* for warm water supply, Blender *21* with Drive *22* for unloading of fermented sediment via Sleeve *23* into worm Thickener *25* with external Drive *24* wherein Screw *28* is installed, and on external side, there are Sleeve *26* for cleared water offtake into Collector *27* and Sleeve *29* for concentrate bend into Collector *30*. There're also on external side of methane Tank *18* Sleeve *31*, connected by Pipeline *32* with Saturator *33* which's linked to Pipeline *42* of purified water supply as well as to Pipeline *34* for supply of biogas solution into Shell *1* via exit Holes *35*.

Being proposed aquaflotoharvester work principle is in the following. Dirty water via Sleeve *4*, and working liquids, respectively, via Sleeve *2* with well-soluble gas, for instance, with carbon dioxide, and via Sleeve *3* with poorly-soluble gas, for instance, air, are supplied inside Shell *1* where combined flotation process of preliminary water purification proceeds. Further process of water polishing goes on in Filter *10*. Then purified water is supplied into intermediate Reservoir *12* wherefrom with the help of Pump *13*, it's given into Pipeline *14* and further is partially outputted by purpose, and partially — along Pipeline *15* for the supply into Saturators *36* and *40* for preparation, respectively, of working liquids of well-soluble gas, for instance, carbon dioxide, and of poorly-soluble one, for instance, air. Such divided supply of working liquids allows to intensify essentially flotation process of water purification. And, together, it's been established on the basis of experiments that the best result appears when working volumes of saturators for preparation of working liquids, respectively, with well-soluble gas and poorly-soluble one in purified water are related as from 1:1 to 1:10, accordingly.

Additional intensification of water purification by flotation is reached also by supply of biogas solution into Saturator *33* into working space of Shell *1*. Combined variant of flotation process speeds up water purification flotation process by about 2–3 times.

Foam product, being formed during realization of combined flotation process of water purification, from foam Gutter *7*, comes via Sleeve *17* into methane Tank *18* for fermentation of mixture of foam product and sediment which are formed in the process of water flotation purification. After fermentation of mixture of the denoted waste, being formed biogas is outputted via Sleeve *31* and then along pipeline is supplied into Saturator *33* whereto purified water is also supplied from Pipeline *15*. Being formed biogas solution is supplied via exit Holes *35*, intensifying flotation process inside Shell *1*.

Fermented sediment is supplied along Pipeline *23* into worm Thickener *2* wherein its thickening goes on which after, thickened concentrate goes into sediment Collector *30*, and cleared water — into Collector *27*.

Quality of obtained clean water, being selected from Collector *12*, satisfies water quality demands of drinking purpose, and sediment thickening degree makes the sediment transportable with residual humidity from 60 to 70%, meanwhile, time of purification process is approximately 1.5–2.0 times less than at the use of known devices.

Being proposed aquaflotoharvetster might be used autonomously qua a local purification installation with the square less than in the case of use of facilities-analogs by up to 2.0–3.0 times.

CONCLUSION

Pursued by us investigations of physical-chemical processes of water purification have shown the importance of usage of kinetic models, including, flotation. Novel approach, based on process multistage nature, has appeared to be successful not only for intensification of processes on existing equipment but also on newly developed one. Together, developing of modelling process has provided for the occurrence of new equipment new type — flotation harvesters which have appeared to be more effective at lower material and energetic expenditures than known machines and apparatuses.

New methods of flotation kinetics and articulated processes, considered in the book, constitute calculating basis for flotation technique, including flotoharvesters, and are already applied in the practice of wastewater purification. The denoted approaches are analogous by formal features to chemical kinetics methods that we underlined yet on the stage of multistage flotation model. Nevertheless, the definition of constants of technological stages for flotation and articulated processes, proceeding in a flotoharvester, is accomplished by other than constants in chemical kinetics methods. Meanwhile, it's worth to note that significant part of mathematical approaches can be equally applied for calculations both, in chemical or flotation kinetics, for instance, as in widely known program complex Maple. The use of such program products allows to embody modelling of rather complex processes, taking place in a flotoharvester. Obtained by us data testify that the developed models make it possible to calculate rather precisely flotoharvester major parameters in view of specifics of being purified wastewaters as well as to calculate regimes of their exploitation.

Accumulated experience of flotoharvester exploitation allows to hope that this kind of water purification equipment would be rather widely demanded already in near time.

The considered question in this work have been developed by us during last thirty years (1987–2020) and, besides, constitute novel matters as for domestic practice as well as for world one [11–12, 50]. Approach, used by us, is based on the modelling of flotation process as in the form of flotation multistage model, taking into account qua major feature a formation of flotocomplexes (hydrophobic particle)-(gas bubble) as in the form of generalized model, taking into account qua major feature a formation of flotocomplexes in the form of mega-air-floccules, containing hydrophobic and hydrophilic particles as well as fine-dispersed gas bubbles. The denoted approach has been checked in its basis experimentally by the author as well as by his pupils, who defended PhD and Second (Russian) PhD dissertations.

To avoid intellectual property loss the author has named the models, developed by him, by his name. In this regard, the author asks developers, who use or will use the author's approach on the description of flotation complexes or mega-air-floccules at the consideration of flotation kinetics, to account for this at the publication of own jobs.

PUBLICATION LIST

1. *Grattoni C., Moosai R.* and *Dawe R.A.* (2003). Photographic Observations Showing Spreading and Non-Spreading of Oil on Gas Bubbles of Relevance to Gas Flotation for Oily Wastewater Cleanup. Colloids and Surfaces A: Physicochemical and Engineering Aspects, 214(1–3), 151–155.
2. *Galbraith P.* and *Stillman G.* (2006). A Framework for Identifying Student Blockages During Transitions in the Modelling Process. ZDM, 38(2), 143–162.
3. *Mostafa N., Syed H.M., Igor S.* and *Andrew G.* (2009). A Study of Melt Flow Analysis of an ABS-Iron Composite in Fused Deposition Modelling Process. Tsinghua Science & Technology, 14, 29–37.
4. *Wills B.A.* and *Finch J.* (2015). Wills' Mineral Processing Technology: An Introduction to the Practical Aspects of Ore Treatment and Mineral Recovery. Butterworth-Heinemann.
5. *Grammatika M.* and *Zimmerman W.B.* (2001). Micro-hydro-dynamics of Flotation Processes in the Sea Surface Layer. Dynamics of Atmospheres and Oceans, 34(2–4), 327–348.
6. *Van der Westhuizen A.P.* and *Deglon D.A.* (2007). Evaluation of Solids Suspension in a Pilot-Scale Mechanical Flotation Cell: The Critical Impeller Speed. Minerals Engineering, 20(3), 233–240.
7. *Szpyrkowicz L.* (2005). Hydrodynamic Effects on the Performance of Electro-Coagulation/Electro-Flotation for the Removal of Dyes from Textile Wastewater. Industrial & Engineering Chemistry Research, 44(20), 7844–7853.
8. *Yianatos J.B.* (2007). Fluid Flow and Kinetic Modelling in Flotation Related Processes: Columns and Mechanically Agitated Cells — a Review. Chemical Engineering Research and Design, 85(12), 1591–1603.
9. *Rodrigues R.T.* and *Rubio J.* (2007). DAF – dissolved Air Flotation: Potential Applications in the Mining and Mineral Processing Industry. International Journal of Mineral Processing, 82(1), 1–13.
10. *Miettinen T., Ralston J.* and *Fornasiero D.* (2010). The Limits of Fine Particle Flotation. Minerals Engineering, 23(5), 420–437.
11. *Ksenofontov B.S.* Flotation Processing of Water, Waste and Soil. M.: Novel Technologies. 2010. P. 272 (*in Russian*).
12. *Ksenofontov B.S.* Water and Soil Purification by Flotation. M.: Novel Technologies, 2004. P 224 (*in Russian*).

13. *Ksenofontov B.S.* Process Modelling of Electroflotation Cleaning of Wastewaters. Express-Information. Series "Industry of Mining-Chemical Raw" NII-TECHEM 1987. No. 4. P. 1–8 (*in Russian*).

14. *Gvozdev V.D., Ksenofontov B.S.* Industrial Wastewater Purification and Sediment Utilization. M.: Chemistry, 1988. P. 112 (*in Russian*).

15. *Ksenofontov B.S., Vinogradov M.S.* Usage of Generalized Flotatotion Model of Ksenofontov for Calculation of Water Purification Processes. Tver: Tver State University. 2019. P. 185 (*in Russian*).

16. *Ksenofontov B.S.* Wastewater Purification: Flotation and Sediment Thickening. M: Chemistry, 1992. P. 144 (*in Russian*).

17. *Deryagin B.V.* Micro-flotation: Water Purification, Enrichment / Deryagin D.V., Dukhin S.S., Rulev S.S. M.: Chemistry, 1986. P. 112 (*in Russian*).

18. *Ksenofontov B.S.* Natural Environment Safety: Biotechnical Bases. M.: Publishing House «Forum»: INFRA-M. 2016. P. 200 (*in Russian*).

19. *Ksenofontov B.S.* Treatment of Wastewater Sediments. M.: INFRA-M. 2019. P. 262 (*in Russian*).

20. *Ksenofontov B.S.* Wastewater Purification Intensification by Flotation. Saarbr cken: LAP LAMBERT, 2012. P. 99.

21. *Ksenofontov B.S.* Water Systems Flotation Treatment. Saarbrücken (Germany): LAP LAMBERT, 2011. P. 189.

22. *Kolesnikov V.A., Ilyin V.I., Kapustin Yu.I., Varaksin S.O., Kisilenko P.N., Kokarev G.A.* Electroflotation Technology for Purification of Wastewater of Industrial Enterprises / Editor Kolesnikov V.A. — M.: Chemistry, 2007. P. 303 (*in Russian*).

23. *Bondareva GM* Development of Flotation Process of Extraction of Surfactants and Motor Fuels from Water Drainages: Dissertation for PhD in Chemistry: Moscow 2010. P. 174 (*in Russian*).

24. *Nazarov M.V., Vineshtock P.N., Voronina A.N.* Preparation of Under-Commodity Waters for Low-Permeability Collectors of Oil by Electroflotation Method // Oil and Gas Business: scientific Electronic Journal. 2014. No. 1. 2014 (*in Russian*).

25. RF Patent on Innovation No. 2108974 «Way of Wastewater Purification», accepted 04/22/1996, registered 04/20/1998. Author and Applier Ksenofontov B.S. (*in Russian*).

26. *Ksenofontov B.S., Senik E.V.* Cleaning of Wastewaters on Flotation Decanters. Life Safety. 2018. No. 5 (209). Pp 21–26 (*in Russian*).

27. *Golman A.M.* Ionic Flotation. M.: Subsoils, 1982. P. 143 (*in Russian*).

28. *Ksenofontov B.S.* Wastewater Flotation Purification. M.: New Technologies. 2003. P. 160 (*in Russian*).

29. *Ksenofontov B.S.* Multistage Model of Flotation and Flotoharvesters: Monography. Tver: Tver State University 2019. P. 194 (*in Russian*).

30. RF Patent on Useful Model No. 183320. Ejector-Mixer, accepted 07/16/2018, registered 09/18/2018. Author and Applier Ksenofontov B.S. (*in Russian*).
31. *Ksenofontov B.S.* Wastewater Purification: Flotation Kinetics and Flotation Harvesters. M.: PH: «Forum»: INFRA-M. 2015. P. 256 (*in Russian*).
32. *Ksenofontov B.S., Kapitonova C.N., Senik E.V.* The Usage of Ksenofontov's Multistage Model in the Processes of Flotation Purification of Wastewaters: Monograph Tver: Tver State University, 2019. P. 162 (*in Russian*).
33. RF Patent on Useful Model No. 170182. Flotation Harvester for Wastewater Purification, accepted 07/25/2016, registered 04/18/2017. Author and Applier Ksenofontov B.S. (*in Russian*).
34. RF Patent No. 2658411. Flotation Harvester for Wastewater Purification, accepted 04/11/2017, registered 06/21/2018. Author and Applier Ksenofontov B.S. (*in Russian*).
35. RF Patent No. 2669842. Flotation Harvester for Wastewater Purification, accepted 11/17/2017, registered 10/16/2018. Author and Applier Ksenofontov B.S. (*in Russian*).
36. *Ksenofontov B.S.* Wastewater Purification in Flotation Columns. Water Purification. 2018, No. 1–2. P. 18–23 (*in Russian*).
37. *Ksenofontov B.S.* Mathematical Models of Complex Articulated Processes in Flotation Harvesters for Wastewater Purification. Water Purification. 2018, No. 10. P. 7–11 (*in Russian*).
38. *Ksenofontov B.S.* Industrial Wastewater Purification from Mineral Oils by Flotation with Out-Extraction of Micro-flotocomplexes. Water Purification. 2018, No. 10. P. 12–18 (*in Russian*).
39. *Ksenofontov B.S.* Intensification of Wastewater Purification with the Use of Combined Flotation Technique. Water Purification 2018, No. 4. P. 8–13 (*in Russian*).
40. *Ksenofontov B.S.* Models of Complex Flotation Processes of Wastewater Purification. Water Purification. 2018. No. 6. P. 59–69 (*in Russian*).
41. *Ksenofontov B.S.* Wastewater Purification: Multistage Flotation Model and Flotoharvesters. Water Purification. 2018, No. 12. P. 5–21 (*in Russian*).
42. *Ksenofontov B.S.* Wastewater Purification with the Use of Ionic Flotation. Water Purification, 2018, No. 6. P. 5–15 (*in Russian*).
43. *Ksenofontov B.S., Stelmach E.S.* Ejectors qua Mixers for Reagent Treatment of Water. Water Purification. 2018, No. 1–2. P. 70–74 (*in Russian*).
44. *Ksenofontov B.S., Stelmach E.S.* Intensification of Wastewater Flotation Purification with the Use of Jet Aerators and Ejectors. Water Purification, 2018, No. 6. P. 25–35 (*in Russian*).
45. *Ksenofontov B.S.* Possibilities of Usage of Pressure-head Flotation Skimmers for Thickening and Inactivation of Active Silt Biomass. Water Purification, 2018, No. 4. P. 14–19 (*in Russian*).

46. *Ksenofontov B.S.* Flotation Method of Active Silt Thickening with the Use of Carbon Dioxide /Ksenofontov B.S., Kozodaev A.S., Dulina L.A. // 7th International Congress «Water: Ecology and Technology» (EQUATEC-2006), V2. M. 2006. P. 830 (*in Russian*).

47. *Ksenofontov B.S.* Water Preparation and Water Offtake. M.: Publishing House «Forum» — INFRA-M. 2018.. P. 298 (*in Russian*).

48. *Ksenofontov B.S., Titov K.* Multistage Ksenofontov Model of Flotation und uts. Saarbrucken: LAP LAMBERT Acad. Publ, 2019. P. 57.

49. *Ksenofontov B.S.* Wastewater Purification: Multistage Flotation Model and Flotation Harvesters. Water Purification, 2018. No. 12. P. 5–21 (*in Russian*).

50. *Ksenofontov B.S.* Generalized Multistage Flotation Model and Development of Flotoharvesters of KBS Type and of Special Purpose for Purification of Water and Soil: Monograph. Tver State University, 2019. P. 104 (*in Russian*).

51. *Ksenofontov B.S.* Multistage Model of Flotation Process for Water Purification. IOP Conference Series: Materials Science and Engineering. 2019. Vol. 492, Issue 1. Art. No 012033. DOI: 10. 1088/1757 -899X/492/1/012033.

52. *Ksenofontov B.S.* Simulation of Wastewater Treatment in Flotation Machine. AIP Conference Proceedings. 2019. Vol. 2195. Art. No 020070. DOI: 10. 1063/1.5140170.

53. RF Patent on Useful Model 123 001. Flotation Machine for Wastewater Purification, accepted 07/24/12, registered 12/20/12. Authors: Ksenofontov B.S., Sazonov D.V. Applier: Bauman State Technical University (*in Russian*).

54. RF Patent on Useful Model 143 014. Flotation Machine for Wastewater Purification, accepted 12/30/13, registered 07/10/14. Authors: Ksenofontov B.S., Petrova E.V., Vinogradova M.S. Applier: Bauman State Technical University (*in Russian*).

55. RF Patent on Useful Model No. 149273. Flotation Machine for Wastewater Purification // Ksenofontov B.S., Antonova E.S., applied 02/24/2014, published 12/27/2014 (*in Russian*).

56. RF Patent on Useful Model No. 192973 "Bioflotoharvester", taken 06/18/2019, registered 10/08/2019. Applier and author Ksenofontov B.S. (*in Russian*).

57. RF Patent on Useful Model No. 194987 "Chemoflotoharvester", taken 08/09/2019, registered 01/10/2020. Applier and author Ksenofontov B.S. (*in Russian*).

58. RF Patent on Useful Model No. 194985 "Electroflotoharvester", taken 08/22/2019, registered 01/10/2020. Applier and author Ksenofontov B.S. (*in Russian*).

59. RF Patent on Useful Model No. 194986 "Deminoflotoharvester", taken 09/16/2019, registered 01/10/2020. Applier and author Ksenofontov B.S. (*in Russian*).

60. RF Patent on Useful Model No. 195481 "Silt-flotoharvester", taken 10/25/2019, registered 01/29/2020. Applier and author Ksenofontov B.S. (*in Russian*).
61. RF Patent on Useful Model No. 195504 "Soil-flotoharvester", taken 11/05/2019, registered 01/29/2020. Applier and author Ksenofontov B.S. (*in Russian*).
62. RF Patent on Useful Model No. 199049 «Aquaflotoharvester», accepted 02/05/2020, registered 08/11/2020. Applier and author Ksenofontov B.S. (*in Russian*).

CONTENTS

www.ingramcontent.com/pod-product-compliance
Lightning Source LLC
Chambersburg PA
CBHW071545210326
41597CB00019B/3126